THE MATRIX OF TRUTH

THE QUANTUM OF REALITY

- BOOK 1 -

DEO Publishing Ltd.

For information about permission to reproduce selections from this book, write to DEO
Publishing Ltd., 5015 W. Misty Willow Lane, Glendale, Arizona, 85310.

Library of Congress Cataloging-in-Publication Data is available.
Library of Congress Control Number: 2012924071
CreateSpace Independent Publishing Platform North Charleston, South Carolina
ISBN 13: 9780988797109
ISBN: 0988797100

Book design by David OLander

Printed in the United States of America
DOC 10 9 8 7 6 5 4 3 2 1

To my son, Caleb, and my daughter, Caris

If I could say anything to you, it would be this: do not strive for wealth, power, ease, or recognition, for these things are remarkably hollow. Be wise in the ways of man, but do not succumb to the foolish things of humanity. Aspire to excellence and always seek the higher ground.

It's the Question

If you are here because of the question, what is truth, reality, purpose, or meaning, then welcome, you are about to embark on the journey of your life.

– TABLE OF CONTENTS –

– INTRODUCTION –

Many times over the course of my life I have been drawn to the deserts of Arizona due to a troubled spirit. This uneasiness has stemmed from what I've witnessed and been exposed to in the world. In the wilderness, one is removed from the distractions of humanity, and the mind is free to assess things from a vantage point of remarkable clarity. During those wilderness times, I have often wondered why the world is such a schizophrenic place and why humanity seems so bent on destroying earth and itself. How is it that we seem to love life, but annihilate it so often? Why have billions been killed at the hands of their fellow men, not only in wars and conflicts, but also by apathy and neglect? I've heard many reasons given for the slaughter, negligence, and the wanton drive for gratification, but these explanations have been superficial at best. In other words, the reasons given for man's ignorance or malevolence rarely, if ever, dig deep enough to discover the root causes, and hence, possible remedies for those problems.

The world is in a much worse position today than it once was—so much so that I fear the earth is now in its death throes while we continue to pursue a course of self-indulgence and self-destruction. Edmund Burke once said, "The only thing necessary for the triumph of evil is for good men to do nothing." We have arrived at this point. Our apathy towards our slide into oblivion has been, and continues to be one of exemplary indifference.

After a roundabout journey and many long years of study, the missing pieces of the puzzle have come together, I no longer guess at the answers to humanity's dilemma. In essence, this series was written to relate what that journey uncovered. The story behind the scenes turned out to be quite as strange as quantum physics itself, the only difference being that one of these forces acts to bind life together (the quantum reality) while the other seeks to destroy it (the linear mindset), an extremely profound difference.

There is evidence enough to be found solely in the environmental destruction of our planet, which is propelling us ever faster towards a total ecosystem collapse (global bifurcation). That fact alone should satisfy the most obtuse critic, but it rarely does. How much evidence has to be presented before a person finally comprehends a truth, simple or otherwise? Apparently there are a lot of people living today who, given any amount of evidence, be it contrary to their worldview, will categorically deny it without so much as a second thought. In an effort to step back and see the entire canvas, we will discuss many forms of evidence, even some forms that you may have rejected, that build and develop a baseline for understanding the problems we face.[1]

There is no doubt that we have entered an era that is unprecedented in the history of humankind. While we glory in our technological and scientific advances, we continue to ignore those non-advances that are killing us. That behavior needs to change if we are to survive. The discussions pertaining to this overall canvas may seem strange, but rest assured, the painting itself is poignant in both its beauty and disfigurement. Beauty, in that humanity is unmatched in its spirit to prevail against all odds, and in its disfigurement in that our spirit to prevail has been many times derailed and usurped. In fact—and quite contrary to our progressive and Enlightenment-based notions of continuous advancement—this canvas shows a regressiveness that is both startling and revelatory in its nature.

---

1 Let's consider a few Socratic questions. Do you think it's a coincidence that more people die of cancer today than have in the preceding history of mankind (as adjusted for population growth)? Do cases of DNA breakdown and chromosomal damage (all passed from generation to generation) have any bearing on what we've allowed into our food supply? Do autoimmune disorders, cell toxicity, and brain damage (i.e., dementia, Alzheimer's disease, attention deficit disorder) have any bearing on what is taking place within industry, government, and the scientific community?

For indeed, the canvas we speak of shows a vanishing landscape. Portions that once conveyed a sense of magnificence and peace have been rubbed out or painted over as if by a child with a Crayola crayon.

Unfortunately, the time to debate the child's artistic abilities is past; time is no longer our ally. If we are to save our planet, we must understand and accept certain truths in order to counteract the poison that has invaded all. If each person does not understand some very basic truths, truths about themselves, humanity as a whole, and the quantum reality that gives life its ultimate value, then any efforts to correct the problems will be misguided and ineffectual, and will eventually culminate in our demise. Death is at the door, my friend; don't treat his arrival as some insignificant wind stirring in the night.

– PREFACE –

I was asked once if my writings were aimed at the general person on the street or at a specific audience. My answer to that question was and still is, why? To categorize this book and the discussions to come would be counterproductive. If I said this book was about science, philosophy, mathematics, history, medicine, sociology, or any other single subject, it would instantly compartmentalize this body of knowledge in a reader's mind. Such is the nature of linear thinking. You will soon notice that this book excludes the finalities of the parts for the embodiment of the whole. You will be introduced to a new way of thinking, one that goes beyond humanity's current linear methodologies to a particular way of quantum thinking.[2]

In many ways, quantum thinking and knowledge makes very little sense to a linear mind. Because of this, I ask you to withhold judgment when parts of our conversations may seem incomprehensible or to conflict. Many of the concepts, facts, theories, and truths presented within this book are not self-evident in and of themselves. The ground that we will be covering is hard, and in many cases, the tools you possess to cultivate it are imprecise or missing altogether. This was precisely my predicament when I started this research project over forty years ago. Today however, the ground has been tilled and is ready to be sown. But herein

---

2 Quantum thinking is the domain of Style VI thinkers. Thinking styles will be discussed in the chapters to come, but for now, suffice it to say that quantum thinkers rarely think in a linear fashion (e.g., A's connection to C is by way of B). A quantum thinker would instead question the meaning and layers associated with A, B, and C, and quite possibly come up with multi-convergent truths previously unrecognized by a linear thinking world.

lies perhaps an even greater obstacle. Will the average person truly comprehend and appreciate the bounty if they have not worked and cultivated the ground themselves? The answer to that question will greatly depend on each individual.

If anything, this book will be about the quantum reality and the Matrix of truth in all subjects, in all forms, in all places, and in all times. That truth is not a personalized truth, but a truth that holds true for all individuals. In other words, the truth we speak of is not reserved for the few; rather, it is meant to be shared with anyone who has an open mind and a willingness to go beyond themselves. That said, our discussions will raise some awkward questions. Why? When dealing with ultimate truth and reality, we must deal with the individual and how he or she thinks and acts. This means things will get personal on many levels, levels that may make one feel exposed where exposure is not desired. But such is the nature of our search for truth: it must be done.

Many people presume (incorrectly) that studies concerning such things as knowledge, reality, and truth deal primarily with the tools of abstract thought, logic, and theory. While those tools are useful, they rarely reach definitive conclusions. Hence, this book is about capturing those loose strings and connecting them in order to discover reality for what it is, not for what it might be. This also means that you will learn about a multitude of disciplines, research, and methodologies in our quest for answers.[3]

If you hold one particular worldview (i.e., political, religious, secular, conservative, liberal, socialist, capitalistic), then there's a good chance we will address it and test it. All systems are fair game in the search for reality and truth. I make no apologies if my findings offend you, for offense is never my motive or goal. Just remember, humanity has existed

---

3   This will include such established systems of analysis as historiography, connected coherence, content analysis, ethnography, convergent systemization or correlational convergence, retrodiction and postdiction inquiry, a priori and a posteriori analytics, inductive and deductive logic, falsifiability examination, inductive coevolution, stratification and network analysis, dialectics, interpretive deconstruction, cognitive analysis, ethical and moral examination, hermeneutics, and phenomenology.

for millennia. Customs, beliefs, and traditions are not only passed from one generation to the next, but also morph and intertwine themselves with each generation's exposure and experiences, creating new trends in thought. Because of this, there is a lot to examine and unwind in looking for the foundational truths that give us life and purpose. If you have already identified yourself with a particular way of thinking or have an exclusive worldview, then whether you know it or not, you have probably chosen to follow someone else's belief system instead of starting from a point of neutrality. This neutral position is the crucial point at which one can find, follow, and act upon the absolute truths of life.

My only true goal is to unlock doors, these being doors for you to consider in your search for meaning. If you want to step beyond those doors, then do so with the full knowledge that what you might find may be disagreeable, but by the same token it could be enlightening and suit your own goals in defining the reality of today. In the end, how you view these truths, interpret them, and use them is entirely up to you.

A little later in this book, we'll spend some time discussing two thinking styles categorized as Styles II and III. These two thinking styles are used by the vast majority of people on the planet. What Style II and III thinkers view as relevant and important forms the basis for many of the world's difficulties today. As never before in the history of the world, every life form and the very planet we call home hangs in the balance. There are many reasons for this, but the thought processes of and actions taken by Style II and III thinkers are some of the primary causes of our problems today. Because of that, we'll spend some time discussing their role as it relates to quantum reality and the linear reality they inhabit.

We no longer have the luxury of making mistakes and remaining confident that life will go on. Even as I write, changes are cascading on a global scale, irreversible transformations directly related to Style II and III thinking methodologies and their collective influence on world affairs. These two thinking styles hold the keys to power and influence over society. These thinkers are the majority, and they need a new revolution in thought and action. Revolutions in thought have toppled governments and countries, to the utter amazement and bewilderment of critics. The only real difference between those revolutions and what faces us today is scale and time. Time is against humanity, and instead of mobilizing the citizens of a nation, we are now tasked with mobilizing the globe. Can it be done?

Many readers will misunderstand or flatly renounce our discussions in an effort to maintain a set way of thinking. In many instances, the mind simply lacks the tools necessary to understand the totality of the situation or the truth, or more than likely, the truth will end up demanding more than the person is willing to give.

Much like those individuals who deny the truth, the individuals who seek it usually fall into three main categories:

- Those who sense there's something more than what's commonly offered up as reality (or the lack thereof).

- Those who have already stepped off the mainstream highway and are looking for the nourishment and the tools needed to continue their journey.

- Those who have a deep desire to change those systems which have proven destructive to life.

Your level of maturity and your attitude towards yourself and others will ultimately determine what you gain from this series. The knowledge it contains will lead to both darkness and light. It's an opportunity for you to discover things you've never known before, to travel beyond the obvious and come face-to-face with the mysteries that link you to humanity, the world, the universe, and ultimately to the infinite reality that stands behind it all.

Some of these discoveries will be astonishingly bright and pleasant, while others will be exceedingly dark and haunting. Some of these unknowns have remained hidden for centuries, even millennia, while others are newly discovered, and still others realized only as a byproduct of our discussions. Some of these mysteries are connected to physics (quantum mechanics, general relativity, and so on), some to biology (our chemical makeup), some to sociology and psychology (how and why humanity thinks and acts the way it does), and some even to law and government (the establishment of the state, rules of precedence, ethics, finance, economic systems, justice, and so on). Together, these connections link directly to such concepts as free will and determinism, and to the new avenues of knowledge and reality that stand beyond our current systems of understanding.

Many arguments, both contemporary and historical, have started from a position that holds a certain agenda. The conclusion was in mind before the evidence was gathered, and it proves nothing more than its circularity; in other words, the argument presupposes its end.[4] Almost every organization today was formed and is maintained by agendas that promote the values of the given entity, but not necessarily the ideals that rest upon ultimate truth. Thus, much of what humanity views as objective truths are in fact subjective belief systems, perhaps holding fragments of truth, giving the appearance of complete truth, but which lack the full embodiment of truth. Because of this, our approach to truth will be a bit different than what most people accept. In this case, we will be taking the long road, methodically following the evidence wherever it leads, starting from scratch and discovering the objective truths as we come upon them. Hopefully, by using this approach, we can separate ourselves from cultural paradigms, political dogmas, nationalistic biases, and our own individualized definitions of reality and truth.

Within these pages you will be exposed to knowledge and concepts that most people will never encounter, much less examine or fully understand. If you allow it, this knowledge can change your life, and it could

---

4 Many a doctoral dissertation has started this way, in which the citations and conclusions prove to be meaningless. The phenomenon is known as *confirmation bias*.

place you in a very unique position within humanity. For example, your sensitivity in all areas of life could grow exponentially from where it now stands. You may even develop the ability to see through people, events, and circumstances in predictive ways. Also, if you allow these truths to fully penetrate your life, you will experience a deep sense of loss even while realizing a corresponding gain. Consequently, you will need to ask yourself two questions. One, can I handle the level of knowledge presented herein? And two, am I mature enough to make the changes this knowledge demands? For once something is revealed and our response to it chosen (to act positively, negatively, or not at all), we immediately become parties responsible to that knowledge. Are you mature enough to see and use this knowledge in light of current trends, and past history, and to use the wisdom that will be necessary to take the next step? Only you can decide.

Be aware that this book is written within the historical and cultural framework of the United States (U.S.). As such, it will use examples relating to U.S. history and the state of affairs in which the United States finds itself today. I was raised and educated in the United States; hence, my extensive understanding of U.S. culture, government, politics, history, and so forth lends itself readily to our discussions. I apologize in advance to readers unfamiliar with the U.S.; however, the examples are just that and should not detract too much from the primary concepts discussed herein. Naturally, the discussions will therefore be based within the framework of Western thought, by which I primarily mean those philosophical systems, i.e. Enlightened concepts and methodologies, that are widely adopted in western Europe and North America.[5] That being said, any search for truth cannot and should not be limited by any one cultural mindset; hence, in many

---

5 Throughout this book, the word *Enlightenment*. will have two distinct meanings. When it is capitalized, it refers directly to those doctrines handed down to us from the Enlightenment period (from the eighteenth to early nineteenth centuries). However, when the word is not capitalized, it refers to the learning process in which we gain knowledge.

instances we shall go beyond westernized thinking and enter regions of cultural norms and biases both unknown and seemingly strange to a westernized society.

Along those same lines, I have endeavored to answer some of the toughest questions in life. For many of you there will either be too little or too much information. The happy medium is a tough thing to find. All I can say is that if there's not enough information for you, then this work will be a good springboard for additional research. Should you find yourself enmeshed in too much detail, then take this opportunity to stretch yourself. You might be surprised at how far you can go!

Another point of possible frustration will be the times when the presented evidence appears to be contradictory, in the sense that some materials or subjects may seem to oppose or even invalidate each other or previous lines of thought. A simple but profound example is encapsulated in the following phrase: *Each of us is different and unique, yet we are all the same.* Contradiction or not? There are many reasons for these relative contradictions, they are as follows:

1. First and foremost, we tend to think linearly. Many of the nonlinear subjects discussed herein will therefore seem to conflict with each other or the linear world in which we live. This is a very tough issue to resolve when dealing with linear thought processes. We grow up thinking linearly, and linear thought processes are constantly reinforced. But linear thinking is an obstacle when trying to reason about things that do not fit linear constructs (i.e., the reality that stands behind us and supports our existence). If we try to reason linearly when we are dealing with nonlinear subjects, there will always be conflicts. If you run into one or more things that appear to contradict one another, I ask you to suspend your judgment until you've read the entire book, then go back to those contradictions and reexamine them in the light of the whole.

2. Secondly, in modernity many of the theories we have come to rely upon and accept as fact (i.e., Newtonian physics, Euclidian geometry, Enlightenment-era paradigms) are so etched upon our minds that any contrary evidence is automatically judged as incompatible and thrown out. If our minds are so circumscribed that we refuse to consider other possibilities, then just about every concept that does not agree with our own will be viewed as a contradiction.

3. Third, one's attitude toward life, other people, and oneself determines a lot of what is viewed as fact. For example, the difference between an ignorant child and an ignorant adult is marginal. Children intuitively know their ignorance, while many adults expressly deny theirs. While a child's attitude is generally one of humility, an adult might have an attitude that is vested in something else. In essence, a person must have the right attitude in order to interpret the information correctly.

4. Fourth, words, thoughts, and statements can be viewed in many different ways, depending on one's cultural heritage, education, and thinking style. Simply put, one word can have multiple meanings for multiple people. If I am trying to convey a concept with words, inevitably there will be people who understand those words differently than I intended. Every reader approaches a subject from different levels of understanding; hence, some things will be more readily apparent to some than to others.

5. Fifth and finally, not everyone possesses the ability to think outside the box. For example, some scientists and mathematicians think very concretely and usually suffer for it, while others use their imaginative abilities to discover new possibilities and new paths to truth. If you cannot think in an imaginative way, many things

will not make sense to you due to seeming contradictions, which will put you at a disadvantage in understanding the totality of our discussions.

One of my many goals in writing this series was to introduce people to thinking and reasoning expansively rather than categorically.[6] Most, if not all, of the apparent contradictions will evaporate when the mind begins to think beyond its linearly built constructs, when we begin to see things not as individual items but as the whole of which they are part.

Now let's talk for a moment about the mechanics of my approach. I'll start with footnoting.

Although this book contains some advanced concepts, I've avoided the academic style of footnoting familiar to research references and PhD dissertations, primarily because it's not conducive for the reader who wishes to establish a connection with the author's inner thoughts. Consequently, when I use footnotes, they are mostly informal, sometimes eclectic, and intended to anticipate questions or to clarify a concept.

Also, I'm a strong believer in the dialectic, the Socratic method of learning.[7] It's a learning style that forces one to think, process, and conceptualize beyond linearized methodologies, as opposed to categorical acceptance of information by calculation and/or demonstration (i.e., as in

---

6 Categorical thinking (a form of linear thinking) is what most of us do: that is, we process information in a linear, intuitive, sequential fashion. Does this mean we are stuck thinking categorically, absolutely not, it just means we will have difficulty when dealing with information that has its domain within the quantum sphere of reality.

7 Dialectic is a dialogue between two or more individuals who may hold opposing views but ultimately desire the truth (i.e., the ultimate truth in all things) and are willing to solve it through rational discourse. This form of rational discovery was used by Plato, Aristotle, Zeno of Elea, Peter Abelard, William of Ockham, G. W. F. Hegel, René Descartes, and many others.

academic footnoting).[8] The primary goal for the reader is to think, not to blindly assimilate, but rather to understand the content from the form.

And lastly, footnotes sometimes distract from, rather than add to, the significance of the undivided whole. For example, if I were to reproduce a spectacular painting within this book and then footnote it with a reference to Rembrandt or Michelangelo, most people would quickly categorize the painting as being superb without studying its whole essence to determine if it's truly magnificent or not. When we place the value of something where it was never meant to be (in this case, upon the artist instead of the work), we end up missing the substrates, the nuances, and the embodiment of its true being (not its categorization, but rather its whole being). So to keep from this distraction, I'll endeavor to focus on the immutability and continuity of the matter and forego the academic approach to footnotes.

This book comprises my own studies, known facts, and specific theories that show the descriptive and dynamic complexity of what we call life, reality, and truth. I took this approach for two reasons. First, as previously noted, we can never hope to achieve complete objectivity by continuing to simplify while using linear methodologies. Second, there is already a great body of knowledge (most of it fragmented) at our fingertips that is ignored or seemingly unrelated in its connectedness to the whole. For example, inductive and deductive reasoning, the very backbone of scientific and mathematical inquiry, combines and synthesizes fragmented pieces of information to create new knowledge. Hence, much of our investigation will follow similar processes, while adding methodologies that address complexity without the use of recursive analysis or reductionism. By going beyond linear forms of knowledge, we'll take the next step into the non-elementary, non-systematized world of dynamic complexity.

---

8 Calculations and demonstrations are a positivist and very linearized approach to understanding reality. In contrast, we are going to discuss the various subjects from the perspective of scientific and philosophical realism, as espoused by Descartes, Carl Jung, Albert Einstein, Erwin Schrödinger, Max Planck, and many more.

- PREFACE -

This work is written in a sequential fashion (yes linearly, already a seeming contradiction in the overall approach, imagine that!). Each chapter builds principles and concepts to be used in the next, why? First, it is because we are raised to think linearly, and secondly, when addressing complex subjects, we must first address them from a linear perspective while at the same time trying to balance the reality of the quantum whole. Due to linear thinking and the complexity of the material, you should be prepared for difficulty and begin by reading each chapter in numerical order. Each explores different threads of reality that will gradually coalesce toward the end of the book. But to understand it from a linear standpoint and then from a quantum perspective, you must learn its component parts in a style you already know and are accustomed with.

Finally, professional and academic works are written in the third person. But in a vein similar to that of René Descartes in his *Meditationes de prima philosophia*, and to the Socratic method of learning, I have purposely chosen the first- and second-person narratives. There are a number of reasons for this. First, it allows you to get into my mind. In other words, you, the reader, will get a much better view of my thought processes via my first-person descriptions. Second, I want to approach the various subjects from a relaxed position. This allows us both to be participants in the journey, and gives the research and discussions a personal touch instead of a detached indifference. To that end, my wish is to engage you on a personal level.

– THE BASICS OF TRUTH: DUALITIES AND REALITY –

In January, particularly in the northern hemisphere, winter winds blow with certainty. Trees have shed their leaves, bearing witness to the icy rains and cold winds that sweep across the land. Everyone hunkers down, preparing for the long year ahead. What will the new year hold? What joys and sorrows await us?

For some reason, January is the time of year when I seem most receptive to listening and learning. This may seem strange to some, but when the winds blow, I always stop and listen. January winds speak plainly of past, present, and future, and they remind me of how small I am in the totality of the whole. But even in this smallness, the winds can and do lift my heart and soul to new heights.

I'm reminded that these winds travel from afar, from distant lands and distant ages. Intellectually I know these winds come from the Pacific Ocean, sometimes from the Sierra Madre in Mexico, and at other times from as far away as the Canadian Rockies, blowing from different regions, but always speaking the same language.

Wind is quantum by its nature: there's no linearity to wind, its world is a non-Euclidian one based in the infinite. We can feel the wind on our skin, but we do not see it; we note only the effects of its passing. From ancient primordial times to the present, the wind has always been. And when the earth exists no longer, those same winds will still be cosmic winds stretching across time and space. Whoever it was that said winds are eternal knew what they were talking about.

Can one fully comprehend the wind without acknowledging the messages hidden within? One can walk proudly against the wind, but what does the wind teach us if not the knowledge of the unseen reality behind our pride? The wind is always a call to listen, to comprehend the incomprehensible. For me, the wind is synonymous with a journey, one of thought and reflection on those things that give us life and happiness. A journey that can take you places that you never thought possible, a journey that would be impossible without those thoughts and reflections. It is this type of journey that we are now beginning.

You will feel yourself tossed about, and probably even get a few bruises along the way, but the journey will be an adventure nonetheless. You will quite possibly end up a different person than you are today. With that said, let's begin, and there's no better place to start than with the subject of truth.

There's a lot of confusion about what *truth* means. Many believe that truth is defined by each individual. Some believe that truth is relative to circumstances and time. And there are those who deny that there's a truth to anything. Since we will be referencing the concept of truth many times throughout our discussions, we need to take a moment to define what it does and doesn't mean.

When looking for the truth, one of the best ways to find it is by examining dualities. Dualities are those absolute truths that will always remain as realities. A common duality with which most of us are familiar is the yin-yang symbol.

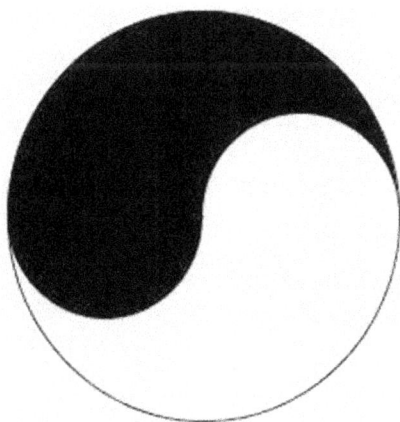

Let's take a moment to look at a few other dualities so as to get the broader picture of how these dualities maintain life, are axiomatic constants, and form the basis upon which objective reality is based:

- We have the concrete reality of a person's <u>determined will</u> weighed against his or her <u>free will</u>.

- We know there are things that are <u>finite</u> in nature; and we also know there are things that are <u>infinite</u> in nature.

- We recognize in physics that light is made up of both <u>particles</u> and <u>waves</u>.

- We grasp the concept that <u>dark and/or black</u> is a state without light and/or white, and the reverse holds true: that <u>light and/or white</u> is a state without dark and/or black. The color grey owes its existence to the dualities of both.

- We know that there are two sexes to our species, <u>male</u> and <u>female</u>.

- We recognize that there is a <u>cause</u>-and-<u>effect</u> relationship operating within our universe; if it were otherwise, all scientific experimentation would be invalid.

- We know that <u>mass</u> is a property of energy and <u>energy</u> is a property of mass (i.e., mass / energy equivalence; $E = mc^2$).

Dualities are those absolute truths that bracket or anchor a spectrum of unknowns.

There is absolutely no doubt that there are grey areas that fall between the various dualities. It is in this grey area of the spectrum that the unknowns and uncertainties exist; the in-between states, those relative and subjective states that are bracketed by universal truths or dualities. The important thing is that in order for these grey areas to exist, they have to have an attachment to or a basis in one of the dualities that comprise the dual nature of the substance or property. In essence, the relative owes its very premise to those absolutes dualities that anchor it. It's always a tricky business to try and deny or redefine an absolute truth by focusing solely on the relative in-between states of the dualities. It's incredibly ironic that more than two thousand years ago Lucretius proclaimed, "There are no absolutes in life." In fact, by the sheer force of his statement, Lucretius nullifies it.[9] We can deny those absolutes, but they always remain as established truths, so that we can have the privilege of denying them in the first place.

Dualities are those objectified truths that bracket what we would consider the unknowns of life. If it were otherwise, all would be chaos, as the anchor points—the universal dualities—would be missing. Some may refute this last statement on a conscious level, but deep down in our subconscious state of being we all operate and function from this imperative, i.e., from the fundamental and absolute dualities in life. To bring home the subject of truth and how we treat it today, we must address some of the best arguments humans have created to refute any possibility of a universal truth or reality. The following examples have had the greatest impacts on modernity today. Hence, we will take a little time to deconstruct them in an effort to determine their validity or lack thereof.

Just one of many disciples espousing Enlightenment beliefs, George Berkeley convinced modernity that individual objects do not have an existence in and of themselves without a mind to comprehend them. In other words, the object only exists in the mind by way of our sensory experiences (via sight, smell, touch, taste, and sound). And since each individual person senses things differently, whether in part or in whole, reality must therefore be interpretive, i.e., constructed by each individual's mind. That

---

9 The statement is an absolute in and of itself. Little did Enlightenment thinkers realize that by denying universal truths, they ended up proving them.

being the case, it logically follows that any ideas of truth and reality must be subjective in nature. An objectified reality or absolute truth would be impossible under Berkeley's worldview. And this is precisely the state of being most Westerners find themselves in today. Because of Berkeley, truth today is considered the subjective domain of each individual person. In other words, each person defines their own version(s) of reality and truth to fit their particular needs, desires, and circumstances.

If you were to examine Berkeley's concept fully, you would automatically see its striking similarities to both Lucretius's statement quoted above and to what is considered the grey areas that fall between the absolutes dualities. Berkeley's argument goes something like this. If a tree falls in a forest, and there's no one present to hear it fall, does the falling tree make a sound when it strikes the ground? According to Berkeley, if there's no one to hear the tree fall, then there's no mind present to interpret the sound; therefore the tree's fall can make no sound. If we focus solely on the in-between areas between the absolute dualities, then Berkeley's ideas of reality would be totally correct. But once we take a step back and examine the absolutes that give life to this in-between state, we at once realize the illusion Berkeley has created.

Our senses are and always have been subjective by nature, but can we honestly make the jump—a literal leap of faith—to conclude that because we are not present to sense an event, that the event produces no effects? Berkeley focuses on the in-between grey areas and never considers those dualities that make the grey areas possible (in this case the dualities of cause and effect). Consider this about Berkeley's argument. Does the idea of matter (e.g., rocks, trees, people) only exist in our minds? From Berkeley's point of view, the idea of matter would only be a state of mind. But since our brains are made up of matter: what does that imply about Berkeley's philosophy? It implies many things, not the least of which is the circularity of Berkeley's reasoning. Absolute reality always produces the experience, without which there would be no experience. A more abstract way of saying the same thing is that the "being" always transcends the appearance. Find out what the "being" is and you'll find the reality behind the appearance. Always be careful of substituting the experience for reality. You will finally know that you have attained true reality when you no longer look to your experiences or those of others as a guiding light and instead use truth for your experience, i.e. action made manifest by truth.

Anything and everything that is not based on absolute reality and universal truth will eventually lead us astray. It is the number one reason why most of our systems are broken today (e.g. governmental systems, economic systems, legal systems, ethical systems, moral systems, etc.).

The drawing below represents a basic linear picture of those elemental parts that allow us to understand universal truth. In reality, all the parts are nonexistent, as they comprise one unified whole. But for the sake of our discussions we will deconstruct the whole to gain a better understanding of its parts and the whole.

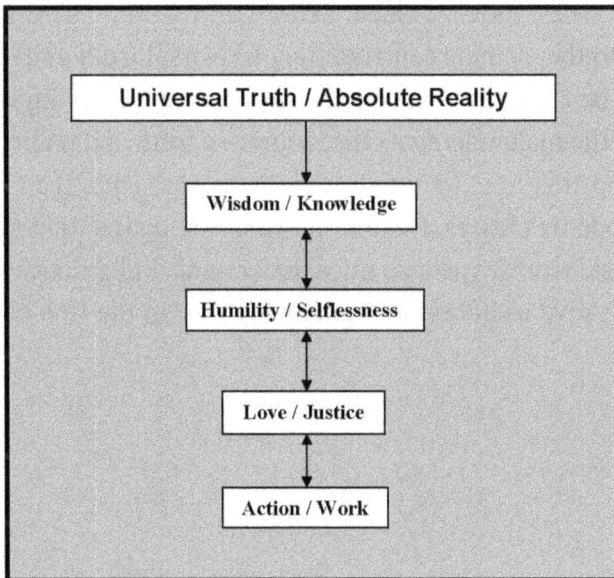

Universal truth and absolute reality have always been the quintessential standard for everything, for things known and unknown as the totality of all. For without the universals and absolutes, nothing has meaning, substance, or life. Therefore, all flows from universal truth and absolute reality.

For example, if my wisdom and knowledge is perfect, then it is 100 percent true and 100 percent real, universal and absolute. If I am a completely humble and selfless individual, then it is 100 percent true and 100 percent real. If I love and am just in all things, then that love and fairness is 100 percent true and 100 percent real. And if I am taking action and working on what I know to be universal and absolute, then the result of those actions and works will be 100 percent true (to truth) and 100 percent real (to reality). Now of course we'll never be 100 percent in anything, at least not consistently 100 percent all of the time. But we can always strive to be better people by working towards the goal of staying within the light of universal truth.

For now, don't get too wrapped up in the concepts of knowledge, humility, selflessness, love, justice, action, and work. There is a lot more complexity to the elements surrounding universal truth and absolute reality than can be captured in a simple drawing. This drawing is meant only to introduce the major elements that comprise truth and reality concerning one's self and this life. As we discuss these elements, their complex natures will become clearer. For the interim, just realize that to understand those universal truths, you also must understand and practice the objective elements that give us access to truth and reality in the first place.

– THE OUTSIDE FORCE ARGUMENT –

One theory that is quite popular today claims to be the final refutation of anything that could be considered an absolute reality or truth. In essence, the theory states that humanity could be controlled by an outside force that gives us the illusion of reality where none actually exists.[10] Given that potential scenario, who's to say that you and I aren't connected to a computer program that feeds us neurological signals that give us the feeling of experiencing life and nothing more? In other words, who is to say that everything we experience is not part of a computer program that creates the illusion of our world and existence? And who's to say that the earth, the stars, and the universe are not illusions created specifically to fool us into believing they're real?

The possibility of an outside force controlling our lives sounds very logical and convincing if we are only looking through a microscope. In this case, we must ask ourselves, what are the dualities or absolutes that give this argument some measure of authenticity? At one end of the spectrum we have the absolute that Descartes articulated so well: "I think, therefore I am." In other words, what can possibly presuppose thinking, the thinker! The thinker (you and I) represents the absolute truth that anchors one end of this duality spectrum.

Likewise, this argument presupposes, and would necessitate, a formal structure (an amalgamation of absolutes) upon which the illusion would need to perpetuate itself upon us, let alone the necessities needed to create the illusion in the first place.[11] Additionally, the argument would have to presuppose the force itself. In this case (being very similar to Descartes's discovery), the structure necessary to create the illusion presupposes the intelligence that created it. That intelligence would have to be real in some form or another. All of this points to the universal truths that must be in place to even suggest an outside force manipulating us in the first place.

---

10 This outside force could be anything that is beyond our purview: an alien society, a god, gods, computer program, etc.

11 If it were otherwise, chaos would reign supreme within the illusion.

Hence, in order to create the illusion, universal truths must first of all exist, the dualities must be real.

To continue this same train of thought, in order for the outside force to create anything, it must begin with something it already possesses. It's this very possession, this reality or truth, that would be needed and would likewise become a part of that which it creates. At least some of the laws that govern the reality of the outside force would have to be used to create the illusion in which we find ourselves. The very fact that we are able to discuss the possibility of an outside force, one that is possibly manipulating us, is proof of something very unique. From a Socratic stand point, what might that unique 'something' be?

In the final analysis, the outside force argument depends on two absolute dualities. At one end of the spectrum sits a human being, and at the other end sits an intelligent being. It's these two dualities that create the grey area in which the outside force argument resides.

So here's the really big question that the outside force argument brings to our attention. Could a person deduce this outside force in a way that circumvents the illusion itself? The answer is a definite yes, and here's why. As human beings, we have the ability for abstract thinking. Since we have the intelligence to think and test abstract ideas, we also have the ability to discern those laws and constructs that emanate from the outside force and its attempted illusion. The one duality, the human "I think, therefore I am," manifests itself in another way, "I reason, therefore I know" (i.e., humanity's capability for abstract deduction).[12] This ability is what sets us apart from the ant that is ignorant of the radio waves that surround him, or the chimpanzee that is oblivious to the neutrinos passing through his body. As humans, we can test for the unseen and deduce those laws that give our world its fundamental necessities. Now, before we leave this topic altogether, I'd like to discuss the uniqueness of the human brain, especially

---

12 Einstein, Schrödinger, and Georg Cantor are but a few examples of humanity's ability to discover truths beyond the obvious, i.e., the illusion or otherwise.

as it relates to thinking and perceiving truths, as this fits nicely with the subject at hand and will be relevant to future discussions.

The wonderful thing about human thinking is its ability to transcend itself. This ability to transcend ourselves gives us insights into areas that would otherwise remain unknown. It helps us to understand others' points of view and enables anyone to dream and imagine something that could be. This transcendence also allows each of us to question our motives and to feel guilt when we have wronged another. This kind of transcendence, combined with the integrative powers of the brain, is nothing less than phenomenal! Now, in anticipation of those who believe otherwise, let's address a misconception that is prevalent today and one that only focuses on those in-between grey areas that we have been discussing.

Scientists can and do break down mechanical processes through the use of experimentation. As such, they have identified regions within the human brain that function as computers, moral regulators, and subconscious filters for emotions and altruistic behaviors. Also, many scientists, medical professionals, and philosophers have defined *reality* as simply the electrical impulses interpreted by the brain. In regards to our senses of sight, smell, hearing, touch, and taste, this definition of reality has some merit. But when it comes to discovering the reality that supports our universe and humanity as a whole, the story is quite different. Let me ask you a question. Does having a biological processor (the human brain) negate the fact that we can produce ideas that are beyond our physical natures? The human mind is the only intelligence (that we know of) that is able to discover the nonintuitive, abstract, and transcendent realities that lie beyond our physical limitations. (Quantum mechanics, general relativity, and dark energy are prime examples of our ability to transcend our biology.) In essence, we are the only creatures that can reason beyond the mechanics of our individual parts. So, should we discount our capability for abstraction and theoretical thought simply because it's contemplated and acted upon by a biological organ? If we used this logic, we would be throwing out some of our greatest discoveries of all time. If anything, the brain being a physical object that is at the same time able to conceive of the immutable should force us to pause and ask, "why?" Why are we capable of producing these extraordinary feats? Why is it that a physical object can comprehend nonphysical realities? The ability to comprehend realities that exist beyond our physical limitations is truly a testament to something. And that something is where some of the deepest riddles in life abide. It has always been, and still is today, those unknown realities of the mind that remind us of our unique and privileged place in the universe. It's what compels us to ask if there's any meaning in life.[13] And this is a solid good that humanity can always cherish and develop!

---

13 This whole question of the mind's ability is significant. Many scientists, medical professionals, and a number of self-proclaimed enlightened individuals limit the human ability to discover truths that transcend biology. Their view is that the mind is nothing more than chemicals and neurons, and it is therefore confined to the physical realities of its makeup. Because of this, these individuals miss much of what it means to step beyond the physical mechanics of our brain to deduce and infer the nonintuitive realities that attest to our extraordinary uniqueness in the universe. It really is time to go beyond our simplistic and regressive models of understanding reality. It's time to embrace complexity with models that incorporate all variables and not just a few.

- TRUTH'S PERSONALITY -

There are parts to truth's personality that are rarely realized. The two that come to mind are cultural bias (e.g., westernized thinking), and the counterintuitive nature of the rule itself. In essence the rule is this: absolute truth is only revealed! Let that sink in for a minute. Then let me say it another way. Absolute truth only reveals itself in proportion to one's capacity to comprehend it, recall the previous drawing on the elemental parts to universal truth and absolute reality? Knowledge and reason alone can only go so far in our ability to understand truth. We need other component parts in order to grasp the truth of anything.[14] And without all the components parts working together, we'll never see clearly enough to understand the fundamental truths of life. It is for this reason that some of the greatest minds in history and today still fail to comprehend reality in its totality. This whole concept is counterintuitive because in westernized societies we are taught that understanding comes by way of knowledge acquisition. It never dawns on the person with a western education that there might be elements missing from their empirical equation. Reality always produces experience; one's experience never produces reality, despite the philosophies and doctrines espoused by the Enlightenment.

Even worse for people taught and trained using westernized methodologies is the fact that much of what we think we know must be unlearned, reassembled, or jettisoned altogether. This is very hard to do when a person stands upon their pride in the knowledge that westernized training gives them. For many Americans, admitting that they may be wrong in a belief or point of view is tantamount to receiving a slap in the face: it's the insult above all insults. As long as we have and hold onto this kind of attitude, we have little to learn and have closed our opportunities to grow in wisdom and maturity. Pride should be the first thing jettisoned when we attempt to learn anything having to do with absolute truth.

---

14 We'll cover these component parts more in depth later in our discussions about Wisdom.

Could anyone actually state that there's no universal reality or absolute truth and expect anyone to believe a statement that purports to be universally true while denying universal truth? Remember, universal truths always define reality. When people start defining their own reality, that's where the real illusions start and the dysfunctions begin. Keep in mind that any justifications given in support of relative truths, whether Berkeley's, the outside force argument, or some other concept, owe their grey musings to the universal truths that bracket and support them. Inevitability and necessitation are those things that cement objective reality and truth. So from this point forward, when I use the words *absolute, universal, objectified,* or any of the other adjectives that describe truth and reality as concrete entities, please know that I'm referring to those absolutes that give life its genuine being and meaning and not to the in-between grey areas leaving life in doubt.

Take a few minutes to study the drawing below. I've used the terms *Relative Truth* and *Absolute Truth* to highlight the paths to each. We will revisit these concepts in later discussions; for now, study the drawing and see what conclusions you can draw and what true enlightenment you may gain from it.

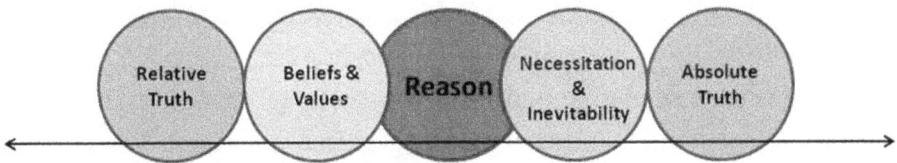

The passion for and the love of truth is what unifies us all. It enables humanity to seek the same ends and to love the same things. Unity, solidarity, harmony, and to live in peace, are these not what we desire and require in life? If we disagree on the fundamental realities of life, we are at war internally with ourselves and externally with the world at large. Universal truths, are the glue that holds everything together.

Wisdom and Knowledge

Let's turn our attention to the subjects of wisdom and knowledge. Since we live in a time and place in which we consume vast amounts of data, let's start with what we know about that knowledge. This is usually referred to as epistemology, the science of how we know what we know.

Knowledge
Human knowledge is made up of four basic parts:

1. We know that we know.

2. We think that we know.

3. We know that we do not know.

4. We do not know that we do not know.

We know that we know - There are things and concepts that we have, if not perfect knowledge of, then at least almost perfect knowledge of. For example, auto mechanics know about automotive engines. They have gone to school to study how the internal combustion engine works, what it is made of, what is needed to make the engine run efficiently, and so forth. They have spent many long hours taking engines apart and putting them back together again. These professionals know that they know auto engines inside and out. There should be nothing, or at least next to nothing that these auto mechanics do not know about automobile engines. Hence, they know that they know!

We think that we know (pseudo knowledge) – This category causes humanity a fair amount of grief. "We think that we know" is a concept deeply embedded in our perceptions of self and the world around us. It is also dysfunctional to learning and to maintaining an open mind to concepts that may compete with our conceptual ideas of what we think we know. There is no doubt that this concept of knowledge is an equal opportunity destroyer of outside ideas and a roadblock to growth and maturity. It has invaded every aspect of our lives and in most cases we don't realize the control it exerts upon us.

"We think that we know" must be considered a pseudo point to knowledge due to the way it employs (either directly or indirectly) the elements of objective truth or reality. In essence, this particular idea of knowledge, when comprehending a cause, theory, ideology, or worldview, has the look and feel of being a reasonable truth, but ends up extrapolating partial truths to the ephemeral world of subjectivity. Hence, this pseudo knowledge is very insidious because most created belief systems start from a seed of absolute truth that gives them a form of validity or claim to truth that the average person has a hard time separating from fiction.

Hence, it is up to the observer to discern if a logical sequence of deductive steps is well-founded and/or justified. Simply put, most people do not want to take the time or effort required to systematically deconstruct complex systems in order to confirm or deny their claims to legitimacy. The examples of pseudo knowledge are more than I could possibly list in this book. I will return to this subject at a later point, specifically when our discussions focus on thinking styles.

We know that we don't know - Examples of this might include knowing that we don't know how electrons travel through space, or how nuclear fission works. The main point to remember concerning this category is that we ourselves are aware that we do not know.

We don't know that we don't know - There are things so distant from our awareness that we are entirely unaware of our lack of knowledge of them. An example might be a member of an isolated aboriginal community who has no concept of spaceflight. But the best way to explain the overall concept is by the analogy of the ant.

The ant knows his world; he knows the trails he walks, the insects he meets, the vegetation around him, his fellow ants, and his enemies. But the ant has no clue about things outside his world. For example, the ant does not know what a cloud is; he does not know anything about solar systems, radio waves, or atoms. In other words, he does not know that he does not know! The ant is well informed about information within his world, but wholly ignorant of knowledge beyond his immediate domain and experience. By his very nature, the ant is limited in what he can know. So what can we infer from this?

It tells us that our knowledge is probably incomplete and will always remain incomplete due to our humanness and position in the universe. As Plato described the prisoners in the cave, our knowledge of most things is partial at best. In many ways, like the ant, we go about our lives unaware of the reality that supports our very existence. Hence, a very obvious point becomes clear. Einstein once said, "We still do not know one thousandth of one percent of what nature could reveal." This whole concept should give us a great deal of humility and caution in how we approach life, specifically in our ideas about reality, knowledge, and truth.

Will we ever be able to grasp objective truths and reality in its entirety? The obvious answer would seem to be no. But are there ways to go beyond our humanness in order to understand a broader picture of reality? The answer to that is a resounding yes, as our future discussions will reveal. Most people miss this opportunity to go beyond themselves due to their preconceived ideas of life (that is, their pseudo knowledge, what we think that we know). If you're an ant, you will think like an ant; but unlike the ant, humans have the ability to step beyond traditional thinking styles to explore and test new and abstract ideas. Einstein noted that imagination was more important than strict adherence to rules. And because of Einstein's ability to imagine, we now have the theory of relativity with all its ramifications and seemingly endless possibilities. So even if complete self-awareness is impossible, our thinking does give us the opportunity to go beyond the typical constraints imposed upon us, unlike the ant.

Wisdom

An Enlightened Recipe for Wisdom's Cupcakes

Start with 2 cups of Enlightenment flour.
Add, 1 cube of anthropomorphic butter.
Add, ¼ cup of experiential sugar.
Add, 1 tablespoon of pragmatic baking soda.
Add, 1 cup of empirical milk (condensed).
Add, 2/3 cups of MBA olive oil.
And finally, add one free-range egg.

Beat until all ingredients are well blended. Pour mixture into an institutional cupcake tin, then bake at 300° Fahrenheit until fully baked to a biased golden brown. Let cool to society's acceptance level, then layer with intellectual icing. When finished, eat and become wise.

This tongue-in-cheek example summarizes how the West defines wisdom. The true recipe for wisdom is quite different from what the world has come to know and expect. For that reason, we are now entering an area far beyond the training we received from Enlightened institutions. In this realm, intellect does not rule, nor does pragmatism, empiricism, or the many other justifications modernity uses to bolster its claims of having wisdom where none exists. When we speak of wisdom, we speak of a complex maze of interrelated parts that in actuality are not parts at all. Such is the difficulty of trying to explain something that is quantumly revealed when our minds want to compartmentalize that which cannot be compartmentalized. Because of this, wisdom cannot be explained in a straightforward, linear fashion. This discussion will actually constitute the first step in moving away from thinking categorically (linearly) and toward thinking quantumly.[15]

---

15 In book 2 we will fully deconstruct wisdom's role.

Today, the whole idea and reality of what wisdom is has been lost. Modern minds believe they know what wisdom is (e.g., a form of reasoned pragmatism seasoned with experience), but in truth, the story is far, far different. Wisdom addresses life on a personal level: the will, the attitude, the intent of the heart. Wisdom is personal in that it's a state of being.[16] While we can do many things considered to be wise acts in and of themselves (such as charitable giving, helping the less fortunate, being fair and just), that is not the same thing as having a state of being that is truly wise.[17] If my will towards myself, others, and life is not right, then I annihilate any hope of possessing true wisdom. In this way, many children are closer to wisdom's door than adults simply because of their innocence.

Wisdom is not having the attributes of truth, knowledge, justice, humility, and love as separate entities, but rather in having internalized them as a unified whole. Another way of looking at this is to realize that each attribute actually uses the other attributes to define itself. Hence, truth means absolutely nothing without the attributes of knowledge, justice, humility, and love defining it. Likewise, love means nothing without the attributes of justice, humility, knowledge, and truth defining it and giving it substance.

---

16 Wisdom may appear to be a type of value system tied to emotions, but while emotions such as compassion and kindness are parts of wisdom's puzzle, they are equally parts of truth's puzzle. In other words, if the emotion is not founded upon or based in a universal truth, then the emotion is subjective at best and a deception at worst.

17 Performing wise acts is always a precursor to adding wisdom to one's being. This is counterintuitive in a culture that equates the acts with who a person is, or the intrinsic character or nature of a person. Anyone can intentionally or accidentally do acts that could be considered wise. But while the act may be good and wise in itself, it does not follow that the person doing the act is a wise person. The confusion derives from how we're educated in today's society. Consider an example from basic algebra: symmetric equality (if $a=b$, then $b=a$). Most of humanity uses variations of this concept when making judgments. But does this idea of symmetric equality have ramifications for how we think and how we apply that thinking to our perceptions of life? Indeed it does, and it becomes an obstacle when determining absolute truths. The main point is that acts of wisdom do not necessarily equal having wisdom (i.e. $a \neq b$), whereas having wisdom necessitates acts of wisdom (i.e. $b=a$).

Wisdom is not obtained by knowledge alone. If we seek it through today's avenues of intellect, then we'll surely miss its true meaning. What we must understand is that wisdom is a perceptual skill and not a formally learned one.[18] Yes, knowledge does play its role in gaining wisdom, but it is by no means the principal player. In other words, if I have attempted to learn what wisdom is from books, teachers, professors, or experience alone, but have not internalized certain attributes within my being (i.e., the attributes of humility, justice, and love), then wisdom becomes unattainable. Hence, I can teach you what wisdom is, by way of our discussions, but if you are missing certain elements, then absolute wisdom will remain beyond your reach. Now, this last statement gives modernity a paradox. In order for me to perceive the universal and absolute truth of anything, I must have wisdom. But in order for me to have wisdom, I must know the absolute truth, and this is a genuine contradiction to the modern linear-thinking mind. However, when we start thinking quantumly, linear contradictions appear as they truly are, half-truths at best [19] The solution, of course, is to start by putting into practice those internal truths that we're born with, namely the ones inclusive of justice, love, and humility.[20] In essence, we start out life with a foundation in absolute truths from which

18 Perceptual in the sense of being perceived by way of having years of experience and not by ways of formalized training. In other words, until a person has experienced certain things, e.g. experiencing death to self, then their mind will not be capable of perceiving the whole (having true wisdom), as opposed to perceiving the parts (having an intellectual understanding).

19 These are created forms of logic that have no basis in reality beyond their ability to imperfectly simplify the complex. As pointed out in the preface, contradictions of the linear sort will be prevalent as we continue our discussions. This is strictly due to the complexity of the subjects and modernity's ingrained ways of reasoning within linear paradigms.

20 A little later in this series we'll specifically address these innate truths, which have been studied and verified by many a researcher in infant and child development. Experiment after experiment has shown that we are born with an innate value of fairness that constitutes the principles of justice, love, and humility. As a side note of equal significance, basic concepts of mathematics are present in infants as well. The blank slate idea espoused by many an Enlightened intellectual—Locke, Rousseau, and Freud, to name a few—turned out to be misguided attempts to explain the world from an Enlightened paradigm.

we begin to perceive what wisdom is by acting upon those truths. As we act upon those universal truths, our mind's develop further understanding of the connections between truths and untruths.

The Five Orders of Knowledge

Let's switch gears now and talk for a moment about the orders of knowledge and how they relate to having wisdom. A point to bear in mind is that any or all of the four access points to knowledge can appear anywhere within the five orders of knowledge. The access points are strictly related to epistemological considerations, while the orders of knowledge are intimately associated with a person's attitude and character development. This is all part and parcel of the complexity of wisdom we spoke of earlier.

The Five Orders or degrees of knowledge are:

- First-Order Knowledge – Having a basic understanding of a closed system environment.[21]

- Second-Order Knowledge – Having an advanced understanding of a closed system environment and a basic understanding of open-system possibilities.

- Third-Order Knowledge – Having an advanced understanding of closed and open systems and a foundation for understanding the connections between them.

- Fourth-Order Knowledge – Having an advanced understanding of closed and open systems and a basic understanding of absolute truths, i.e., objective reality, ultimate meaning, and the roles that humility and wisdom play in life.

---

21 Refer to the glossary for definitions of closed and open systems.

- <u>Fifth-Order Knowledge</u> – Having advanced understanding of closed and open systems, universal truths, objective reality, ultimate meaning, and wisdom, and a basic yet clear understanding of the Matrix of Life.

As a person progresses from one order to the next, a sense of their responsibility to others and the world slowly emerges. Building upon their wisdom, they comprehend and apply the true nature of objective reality. A person moves from being more self-focused to more other-focused, from looking inward (to self-interest) to looking outward (for the greater good). As they move through the third, fourth, and fifth orders of knowledge, the person no longer seeks to be understood as much as they seek to understand. Personal actions and behaviors become ever more deliberate. By the time an individual reaches the fourth and fifth orders of knowledge, they live a life of extreme simplicity (in some cases poverty), and intense service to the greater good of the planet and humanity become prime objectives in life. Also, individuals who achieve the fourth and fifth degrees of knowledge come to realize the nonlinearity of existence, understanding that many manmade definitions are limited to the point of being nondefinitions. For example, separating the ideas or definitions of words such as *morals, justice,* and *love* is absurd when we realize that each of these words is rudderless without the other. These words infuse each other; all are in fact synonymous in that each must make up the definition of the other in order for us to realize the totality and the reality they convey.

One final note for students who wish to challenge their minds a bit further: Diogenes once said, "I am neither Greek nor Athenian, for I am a citizen of the world!" When one finally understands the objective reality that supports the whole, definitions that once applied to ourselves (I am an African, an Asian, an Indian, a democrat, a republican, an atheist, a theist, an American, a European, and so on.) begin to lose their meaning. Why might this be? First, if a person finds too much of their identity within a group, an organization, or a nation, they lose the ability to see things from an objective standpoint. In other words, manmade definitions for groups and organizations always place limits upon objectified truths in an effort to legitimize and perpetuate the group's chosen belief system(s). Second, because the modern world is rapidly becoming one organism by virtue of technology, interlinked financial and economic systems, and the population explosion, we can no longer afford segregationist attitudes identified with traditional definitions of groups. In essence, as you advance in knowledge and wisdom, your perspectives on life will look more and more like that of Diogenes. You will come to realize that truth is not necessarily found in one particular cause or system. It is found in examining and testing all systems while standing outside of them, not within them. As you learn more, you will become aware that your story is part of an immense drama and that each of us plays a part in it, for good or ill. No one is exempt; nothing is empty of meaning.[22] So, do we chose Diogenes' attitude of being a citizen to a greater good (i.e., universal truths and absolute reality), or do we remain forever stuck in polarized systems that narrowly define reality in an effort to perpetuate self-interest?

Socrates was sentenced to death for explaining truths that stood beyond the subjective beliefs and systems of his day. Jesus was crucified for exposing the hypocrisy of pride. Gandhi and Martin Luther King Jr. were murdered for trying to separate prejudice from the truth. Each understood the greater reality and the higher truth, each understood the dangers of narrowing alliances to certain systems, organizations, or groups that obscured those objective truths, and each realized the futility of inaction in the face of injustice and immoral insistence. Wisdom sometimes does not seem very wise in a world that is worried about image and self-preservation, but is it really unwise?

We could stop there, and even jump to the conclusion that humanity is doomed to destruction because of its pride and intractable biases. But there have been instances in history where the masses have come together in support of universal truths in direct opposition to those systems used

---

22 Irenaeus

to justify power and the status quo. Buddha, Saint Francis, Confucius, Martin Luther, and William Wilberforce not only survived their efforts to bring objective reality to the forefront, but also succeeded in persuading and enlightening the masses to take action specifically because those universal truths were brought to light. Humanity does have the capacity to rise above those things that make life unjust and untrue.

– THE MATRIX OF LIFE,
AND THE QUANTUM HOLE –

"The beginning of eternity, the end of time and space,
The beginning of every end, and the end of every place".

Lord Byron

Streaming through the windowpanes, the yellow glow of a fading sun filled the room. The room and house had long since been abandoned, quiet now, left to nature's whims. It was a place I liked to visit. Sometimes I would take a stroll in the nearby woods or fields, and other times I would sit on the aging front porch, listening to the sounds of life that filled this special place - the pulsating beat of a woodpecker, a mouse scurrying through dry leaves, trees groaning in stormy winds, the patter of gentle rain on forest leaves, and often on quiet days, the electric songs of hummingbirds high in the trees. All of those sounds made for a perfect day.

In this house, I had a favorite room. It was an upstairs bedroom with a high ceiling, tongue-and-groove pine flooring, and a window view of what was once a well-maintained garden. The garden had long since gone wild, but it was still beautiful in its own way. In the summer months white hollyhocks, orange coneflowers, purple cosmos, blazing red fireweed, and a whole host of blooms fought for their positions among the weeds and saplings. During the winters, when the trees were bare, I could look through the branches to where a stream flowed into a pond. And when winter winds blew, the pond always froze while the earth dreamed of warmer days.

One could always tell when winter was losing control of the place. When white and purple crocuses pushed their way through the frozen ground, winter was dying of old age. Soon golden daffodils and red tulips followed, and before you knew it, spring appeared with summer on its heels. In all seasons, the old homestead was a wonderful place to visit, a magical place where I always learned something new.

I can still recall one summer's day in particular when I was in the upstairs bedroom, sitting on the floor. As I leaned back against the far wall, the window seized my attention. As the sun set, the softness of the light illuminated small particles of dust floating in the air. Each tiny prism refracted light like heaven's stars. Tiny flashes of gold, blue, silver, and red played upon the invisible streams of light. Time seemed to stop. Every now and then, I'd see some small particle zip from the shadows and streak across the stream of sunlight, only to disappear again between the borders of light and dark. I imagine it was some infinitesimally small insect zipping from place to place, but it appeared to me as a loosed electron hurrying to catch its ride through time and space. As one reflective particle appeared in the light, another would disappear in the dark. Slender streams of the sun's fading light cut through the shadows and slowly worked their way across the floor. Dappled light played on the floor as the leaves outside vied for the sun's attention. Particles that were usually unseen were now seen, and my mind was filled with wonderment and thoughts of the deepest kind.

I reflected in those final moments, in the in-between state where neither light nor darkness prevailed. There is a splendid beauty in that place where the two somehow become one in a mysterious union, light and dark. It's in moments like these that I truly live. Indeed, these things give life depth and value! Such small things that go unnoticed by most, but they are in fact the ocean tides that inevitably draw all to the source of life.

In those final minutes before the light was lost completely, I could still see those small particles suspended, slowly swirling, so small, but very much a part of a larger whole. It was amazing to contemplate.

The realm we'll enter next is one of complexity and mystery. We will be addressing the Matrix of Life (the Matrix) and how it fits into the reality in which we live. The Matrix includes many things, among which quantum mechanics plays a central role. The proverbial question is why? Why indeed, to understand quantum mechanics is to understand the Matrix of Life at its most foundational level. For this reason, we'll be spending quite a bit of time discussing its realities, truths, and ramifications for humanity.

It's amazing that almost a hundred years has come and gone since quantum mechanics was discovered, and yet we still do not understand it. Quantum physics seems to be the great dark void of our time, a place where simultaneity, superposition, entanglement, and the infinite all reign supreme. Even more amazing is the fact that many of these quantum wonders have been established as objective truths and yet remain beyond the understanding of most. It's a bizarre and counterintuitive world that causes us to reassess our notions of reality and the very meaning of life itself. Quantum mechanics explains much of what the Matrix of Life is, so much so that by coming to know the one, we can come to know the other.

Much of our discussion on the Matrix deals with a fair amount of science, primarily physics and cosmology. Thus, some concepts and terms may be unfamiliar if you do not follow the latest scientific news. If you are in that category, this will be an excellent opportunity to stretch your mind and do some independent research.

Most of the concepts mentioned below have been verified through scientific inquiry or have been inferred (deduced) as a result of scientific inquiry or experimentation. If a concept has not been proven or deduced by modern science, then I will state as much so you can differentiate between the various sources of information.[23] With these clarifications in mind, we move into the very real and yet bizarre world that connects the finite to the infinite.

Around the fifth century BC, Democritus and Leucippus suggested that our world was made of absolute and unseen particles called atoms. Plato and Aristotle postulated that all matter had some underlying and hidden form that supported life. What is truly astonishing is that these

---

23 Due to the nature of the subject and our incomplete understanding of quantum mechanics, deductive reasoning and inductive logic will be our primary tools for discovery. In many instances this is where science overlaps with philosophy, the finite with the infinite, and our closed system with the open system that surrounds us. The glossary in the back of this book will help you to understand the various concepts and technical points.

philosophers, using reason alone, were able to predict discoveries of modern physics.[24] This is a key point to remember as we go forward, for much of our discussion will deal with discoveries found by way of deduction and not by empirical methodologies.

By observing cause-and-effect relationships via experiments, modern science has confirmed that energy infuses everything in the universe. We still do not know the true nature of this energy, but we do have a pretty firm idea that it's universal in nature, unseen, and somehow connects and touches everything. In essence, this infinite field of energy is all about the transfer of information. Particles that are light-years apart and yet communicate instantly with each other defy our current understanding of physical laws (i.e., nonlocality, superposition). Physicists refer to this interconnectedness as *entangled states* or simply *entanglement*. Hence, it would seem, at least on the atomic and subatomic levels, that there is no such thing as separation of time and space. All seems to be connected as if it were one entity. Thus, quantum mechanics implies a vast matrix of connectivity within the universe that connects particles to particles and parts to parts, even across light-years of time and space. This instantaneous communication and connectivity between particles, distant galaxies, and solar systems is nothing less than phenomenal![25]

Beyond that, there's also the actuality of infinite space and infinite time. According to quantum mechanics, a radioactive atom can and does exist in two simultaneous states at the same time: it's considered to be both decayed and not decayed simultaneously. Physicists call this phenomenon *superposition*. In a strictly linear, closed system world (i.e., the world of classical physics), we know of no such possibility. When we physically look at or measure an object, that object always appears to be in a definite state (either in one state of being or another). So, from a

---

24 This illustrates another example of the mind's ability to deduce and anticipate the unseen and the unknown (Rationalism).

25 Bear in mind that large and massive objects are made up of individual particles (atoms), so this interconnectedness applies as much to the macro world of planets, stars, and galaxies as it does to the subatomic micro world of atoms, electrons, and photons.

linear, closed system perspective, while we can observe the atom as either decayed or not decayed, we never observe it as both. Another way to visualize this concept is to reflect on the attributes of water. Water can be in many states (e.g., vapor, liquid, solid), but we never see it in all states at once. But in the quantum world, there are no separate states of being, and as bizarre as this may seem, it's already been proven to be the case on a quantum level.

In the mid-1930s, Erwin Schrödinger gave humanity a famous thought experiment, dubbed "Schrödinger's cat." The irony of Schrödinger's thought experiment is that it was originally conceived to show the incompatibility of quantum mechanics with classical states of being (i.e. the incompatibility of a closed system with an open system), but it has turned out to be quite the opposite. Schrödinger conceived of a box that was completely sealed off to the world in which we live. The inside of the box represented the open system where superposition and simultaneity exist together (a quantum or open reality). Schrödinger further postulated that a live cat, a radioactive object, a Geiger counter, and a capsule containing poisonous gas would be placed in this box. A mechanism would be connected to the Geiger counter and to the poison capsule in such a way that if an atom from the radioactive object decayed, the Geiger counter would detect it and trigger the mechanism that in turn would release of the poisonous gas. Of course, once the poisonous gas was released, the cat would die. So what's the significance of this thought experiment, and how is it related to reality?

If quantum mechanics prevails within the box, the atom from the radioactive object will be both decayed and not decayed at the same time: it stays in superposition as long as it remains within that open system within the box. So, does the cat live or die? If the atom decays, the cat dies. If the atom does not decay, the cat lives. If the atom is both decayed and not decayed at the same time, then the cat is held in a state of superposition, neither alive nor dead. The cat exists in a state of equilibrium or coherence, a state that is removed from our classical, temporal world and yet still maintains a vital connection to it.[26] Is your mind connecting the dots yet? Think on this. The cat would be neither dead nor alive because there would be no such thing as time. Time has to elapse for the atom to go from a pure state to a decayed state. With time removed from the equation, the atom is neither pure nor decayed, it just is! It exists in an infinite state of being. Why, then, do we not observe things in infinite states of being on earth? Why indeed!

Objects in an open system, such as the inside of Schrödinger's box, are complete, balanced, and self-contained. The system has all the information it needs to operate efficiently and effectively.[27] And although we are somehow and mysteriously connected to the open system, we/planet earth, still do not carry many of the parts that constitute a fully open system. We do not live in a open system even though we are maintained and surrounded by an open system. But because we are connected to an open system, we can still observe the interactions between the two systems as you might observe the sun through an infrared filter.

To take this analogy further, let's say you place this infrared filter between you and the sun. Photons coming from the sun will be disrupted by the filter. When this happens, their particle waves are reflected and then collapse; phase angles in energy and even time are interrupted and displaced. Hence, much of the information that was included in the sun's initial rays has now been disordered and excluded by means of the infrared filter. From a closed system perspective, this is exactly what happens when we try to view or measure quantum states that reside within an open system. In other words, information is lost or leaked, and it is this

---

26 Some physicists refer to the collapse of a wave function as being the end of a coherent state. It should be noted that the wave function used in physics is still considered a function of space, momentum, and in many instances, time as well. The wave function is a mathematical expression of what many believe superposition to be. It's also a reductionist approach used to remove the many complex interactions involved in quantum mechanics. In essence, it's done to delineate, quantify, and make linear something that is not.

27 Information in this context refers to what is exchanged between particles (i.e., the information instantaneously shared between particles in the quantum world, known as entangled states).

information interference that causes us to see things in their partial states instead of their whole open-system states. Because of this, the totality of reality is partially blocked and concealed behind a filter, that filter being our closed system. In physics, this blocking, disruption, and leakage of information is known as decoherence. Decoherence is the loss of balance (coherence) within the open system. When too much of the information within the open system is leaked or disrupted, it creates a system other than itself, and that system just happens to be a closed system, the very system in which we live and breathe.

Many physicists have come to the conclusion that large objects such as planets and stars—or cats, for that matter—are excluded from the quantum world of simultaneity, entanglement, and superposition, why? Small things, such as particles, have been shown to exhibit these quantum structures, whereas larger objects have not. There is a problem with this line of thinking however, for all things, no matter their size or shape, are made of particles. And if the smallest particles operate within quantum states of being, what does that imply about the objects those particles make up? To find out, let's return to Schrödinger's cat for a minute.

If we were to crack open Schrödinger's box, the closed system that exists outside the box would disrupt the open system that exists within the box.[28] When the box is opened, the information that kept perfect balance within the box comes into contact with our closed system filter. In this case, the filter happens to be us (the observers). The phase angles, the wave functions, and all of the information that was once contained and balanced within the box is now exposed to our closed system filter(s). Looking through this closed system filter we see things in certain arrangements, positions, and defined states of being (i.e., the cat is either alive or dead, but not both). In other words, we don't see

---

28 Schrödinger's thought experiment is a bit confusing in that a closed box represents his open system and everything that surrounds the box represents the closed system. Logically, the box should be analogous to a closed system since it is smaller than the open system which surrounds it.

things as they truly are; instead we see the objects in a decohered or collapsed atypical state of being. So let me ask you: From a closed system perspective, is the cat dead or alive when you open the box? Be slow in answering this question, because it requires more than the obvious answer. It's truly the quintessential question surrounding much of what we view as reality!

Let's pause for a moment. We've covered a lot of ground in a short time and probably need to sit back to digest a few things. When you've had time to think things over and want to bend your noodle a bit further, consider these observations and questions:

- In quantum mechanics, there is no such substance as space and no such thing as time; these mechanisms only apply within a closed system. Since our closed system is surrounded by, sustained by, and operated upon by an open system, are we really detached from it? Do we really live in a closed system, or is the closed system part of the open system? Albert Einstein once said, "…people like us, who believe in physics, know that the distinction between the past, present, and future is only a stubbornly persistent illusion."

- Quantum mechanics applies on all scales, micro as well as macro. We know this because everything is made up of particles. Because we know this, the reality we know and don't know is an absolute quantum reality. It's a reality beyond the classic models of time and space. In quantum mechanics, particles communicate instantaneously from hundreds and thousands of light-years away (i.e., simultaneity, entangled states of being). This is all possible because time and space are irrelevant; they don't exist in an open system. So how does humanity respond once they discover that life is but a placeholder emerging from a spaceless, timeless, and infinite reality?

- The quantum reality that surrounds us, that is in us, and sustains us, is an indivisible whole. It literally is the embodiment upon which being is defined.

- We do not physically see quantum mechanics at work due to two primary factors:

   1) Our worldview, tradition, and persistent faith in what our eyes tell us (that is, the Newtonian model of physics and the Enlightenment model of reality), and

   2) The phenomenon known as decoherence, the disruption and leakage of information from the open system to the closed system.

- The larger the object, the more susceptible it is to decoherence (this includes planets, galaxies, humans, and animals). Likewise, the smaller the object, the less likely decoherence is to take place (i.e., in the subatomic world), and the more we can slow or stop the information leakage in order to observe quantum states. If we can slow or stop quantum states long enough to study them (i.e., to focus on what we can control for, in this case the minute), and to know of their existence, then what are the implications for the larger objects in the universe?

- If quantum mechanics explains much of what could be considered absolute reality, the question is no longer "Is the cat dead or alive?" The question moves from the black-and-white world of our closed system to the full spectrum of the open system. Far from the cat being alive or dead, the cat now exists in an infinite state of being. There is no such state as "dead" or "alive" in an infinite state where space and time do not exist. So if a person dies in a closed system, are they really dead in the open system? One wonders!

• Where does the energy and information contained within each particle and the entangled state of our existence originate from? In other words, what is the source that keeps particles interacting and communicating as a single entity?

– DIVING DEEPER –

"Deep into that darkness peering, long I stood there wondering, fearing, Doubting, dreaming dreams no mortal ever dared to dream before".

Edgar Allen Poe

This timeless, unseen, and unexplainable entanglement within our universe is but a part of a much larger web of connections called the Matrix. The Matrix is the definite reality that connects the finite with the infinite and our closed system with the open system. It's the entanglement of the seen and unseen, of matter and antimatter, of energy, information, and communication. It's the entanglement of knowledge and information with matter and substance.[29] The Matrix is the absolute unity that underlies and supports life as we know it.[30] Thus, everything derives its being and meaning by way of the Matrix.

As previously mentioned, we know that the Matrix is composed of energy and information that is observed in our closed system as well as that which is unobserved in the open system. It includes gravitational energy, nuclear energy, mechanical energy, chemical energy, electrical energy, dark energy, and energy that is both potential and kinetic. It is energy that is made up of specific parts (e.g., electrons, photons), energy that is composed of progressive or developmental parts (e.g., chemical reactions, compressional and informational waves and/or streams), and energy and matter for which there is no clear understanding of the hidden parts (e.g., dark energy, dark matter, the fine-structure constant.).

---

29 From a linear perspective, there is a great chasm between the nonphysical world of thoughts/ideas/information and that of the physical world. But when one examines the supposed gap between this intangible world and the physical world of matter, one sees that each has direct links to the other. In other words, the physical world could not exist without the transmission of information/knowledge between its various component parts.

30 Life is the universal building block of everything we hold dear. Life is not simply a biological happening, it's an organized system of aligned interest and interdependencies. The biological component of life is but a part of a much larger whole.

The energy contained in the Matrix is the workhorse of everything and for everything. It fuels our bodies via energy stored in plants and animals; it allows our brains to process information via energy and information stored chemically and electrically; it allows our planet to create ozone, carbon, and heat as energy stored in the bonds of molecules and atoms; and it allows our universe to operate as it does, through energy stored in the motion of light and mass ($E=mc^2$), in distance and force ($F=mg$), and in matter and antimatter (e, e^-).

Along with the energy component of the Matrix, we also know that it is infinite and transcendent in nature. We know this in part via the links between quantum mechanics, general relativity, and the arrow of time. As it turns out, time is the superhighway that connects our temporal reality (the closed system) to the transcendent reality (the open system). For instance, humans fully experience the arrow of time. We can observe trees growing from saplings; we see people growing older and experience it ourselves; and we view our universe as it continually expands. The arrow of time moves in one direction, from a beginning to an ending. This idea is not restricted to an abstract world. Physicists today grapple with the arrow of time and the problems it poses to many a theory. The arrow of time testifies to the validity and temporal nature of our finite existence within a closed system, and it also testifies to something quite extraordinary.

If you were to follow the arrow of time in a backward direction, you would be traveling back through time itself. You would travel back to when the Romans were the supreme rulers of the known world, to the time of the Clovis people, to the time before humans made their first appearance on the planet, and eventually to the very point where our universe (the closed system) began: time Alpha, the Big Bang (BB). As far as we're able to determine, through physics, cosmology, and mathematics, our universe started out with some form of matter and energy. From a strictly deductive perspective, this starting bundle of matter and energy had to have been compressed to the subatomic level. In essence, every bit of energy and matter that comprises our universe today was initially compressed to a size smaller than anything we can see with the naked eye. The spot where all this compression of energy and matter took place becomes the unifying point at which quantum mechanics and relativity merge. It's that quintessential moment, just prior to the explosive birth of our universe, that the infinite manifested itself in the finite. It's where the arrow of time began; it's the dividing line between an infinite reality and a finite reality, between the open system and the closed system. To understand this concept better, let's drill down even further.

Currently the scientific community is at a loss when trying to combine quantum mechanics with general relativity. Science knows the two systems exist and work together; they just don't know how they do it. And the reason they don't know how it's done is due to some of the same reasons that Einstein could not initially unite Newtonian physics with his theory of general relativity: he was biased toward the traditional norms of his colleagues.

Einstein was trapped within the paradigms of his age, just as physicists today are trapped within the paradigms of our age. In Einstein's day, he joined Newtonian physics with general relativity by creating an artificial link, a mathematical construct that Einstein dubbed the "cosmological constant."[31] Likewise, physicists today attempt to unite quantum mechanics with general relativity by creating their own artificial links (e.g., string theory, loop gravity, and other mathematically exotic constructs) where the mathematics are manipulated to fit within the accepted paradigm (i.e., confirmation bias). For example, when physicists today try to unite the fields of quantum mechanics and general relativity, they are frustrated in their attempts to make the mathematics of the two theories work together. Why? The short answer is that when the two theories are mathematically combined, the answer is always infinity. And, as all good physicists know, when an equation yields an infinite answer, then something must be wrong with either the theory or the mathematics supporting it.

---

31 This is not to be confused with the cosmological constant we use today. The two are very different, please refer to the glossary for the difference.

Now let's look at the long answer.

The apparent trouble exists with our concepts of gravity. General relativity and quantum mechanics both deal in approximations. Mathematics and science use approximations due to the extreme complexities involved and the overwhelming desire to simplify everything to some form of measurement. The issue of infinity is emblematic of when physicists routinely sidestep what they don't understand or can't measure by manipulating the concept, theory, or the math to yield linear or finite results where none exist.[32] Hence, when combining the nuclear forces and the electromagnetism of quantum mechanics with Einstein's gravity, one ends up with a never-ending series of exponential powers multiplied by Newton's constant.[33] Immediately, one can see the problem. The letters in the equation (a, b, c) are unknowns: they are pure numbers, dimensionless placeholders whose future values will never be known with any certainty. Since these pure numbers are infinite in nature, there can never be a finite answer to the quantum / relativity unity question because they will always yield the infinite (primarily due to the fluctuations in quantum geometry and space-time dimension). Now with that in mind, let's return to how this relates to the Matrix.

The arrow of time has its origin in the BB. As far as we can extrapolate, time did not exist prior to the BB, but rather was a result of it. Now here's where your mind will be stretched yet again. Quantum mechanics and general relativity come together and meet at a point smaller than the Planck length, where all the mass and energy in the universe is smashed together to an infinitesimally small point (here are our contradictions and dualities again, something massive, yet something small). Our minds have a hard time comprehending the reality of such a place. But let's ask the question: if the point is so small we can't measure it, and if it is the place where time has not yet begun, and if the mass and energy of this point are both

---

32 The Cartesian plane is a classic example.

33 The strength of gravity (G), times even powers of the energy of what's being approximated (energy or mass, E). For example, $(+aGE^2 +bG^2E^4 +cG^3E^6 + ...\infty)$.

small and large at the same instant (superposition), so as to defy measurements, would not that point, or that space, or that place, or that reality be an infinite field of being? Similarly, if we were to think about time dilation, and we noticed that time had stopped while we were zipping across the universe, what would we call that space, that place, and that reality where time ceased to exist? We can conceive of that which is both timeless and spaceless only in the infinite dimensions of an open system, via deduction. The truth is that infinity exists. The arrow of time starts from an infinite state and leads us to an objective reality, i.e. the passing of time, as attested to by the arrow of time. It's always the infinite that supports the finite (i.e., the open system sustains the closed system), for without the foundation of quantum bits of information and energy, nothing in our closed system would exist.[34] But several questions confront us. Where did the energy that produced the big bang come from? Do scientific laws like the conservation of mass and energy apply in an open system?[35]

If we use current methodologies in science (i.e. cause-and-effect reasoning/relationships), we could very well make the case that something had to exist prior to the big bang.[36] From this point there are basically two paths one could follow. The "loop" path states that energy and matter had

---

34 The BB was created out of pure energy (Einstein's mass/energy equivalence, e.g., $E = mc^2$). We know this because energy is required for work. Gravitational fields, subatomic particles, atoms, and so forth need energy to do what they do best, work! No energy means no work and consequently no BB.

35 As a caveat, in some instances matter can seem to appear from either energy fields or from absolutely nothing (via fluctuating energy fields and/or vacuum fluctuations). This whole concept exposes more questions than answers. Does energy beget matter? Does matter beget energy? Are the terms *matter* and *energy* misnomers for something that is a single entity? Could something really appear out of absolutely nothing, given the concept of nothing?

36 The bread and butter of modern science deals in two very deterministic realms. The first is the realm of causes and actions (i.e., the bread), and the second is the realm of results and effects (i.e., the butter). In other words, scientific proof is only known by humanity according to the effects of the actions. This cause-and-effect relationship not only applies to the physical sciences, but is also equally valid in the social sciences as well. This is an important point to remember when considering social issues.

to exist prior to the big bang, due to the law of conservation of mass and energy. In essence, according to the laws of physics, everything that exists now has always existed in one form or another. The "something from nothing" path instead asks the question, Why? Leibniz asked, "Why is there something rather than nothing?"

To my mind, the loop path is the easier one to follow from a linear perspective. It does not require a lot of imagination or abstraction; it follows the axiom that simple is best (the law of parsimony) and only compels us to answer the hows of the circumstances. On the other hand, the something from nothing path is complex. It requires us to ask why, and it compels us to think beyond the loop (i.e., Was there an actual beginning?). For the most part, the whole concept of beginnings opens the door to possibilities beyond the physical dimensions of a looped system. And because of the complexities involved, most people avoid asking the whys and simply settle for the more simplistic ways of asking how.

Contrary to what you might guess (since I'm usually drawn more to the complex than the simple), the loop system seems to me to be the more reasonable, for the following two reasons:

- Cause-and-effect relationships appear to be universal truths that apply to both closed and open systems.

- The law of conservation of mass and energy would suggest that *something* was there before our universe was born. One can't get something from nothing (again, a cause-and-effect relationship). That something was probably nothing less than pure energy. If we think of the loop system, that energy has always been there. In other words, it would be an energy source that has no beginning and no ending; it just is and has always been.

This second point opens up some interesting debates. It assumes that matter and/or energy were always there and negates the causality of that matter and/or energy in the first place. It becomes a pretty big contradiction to overcome. Can we accept a final verdict? Which came first, the chicken or the egg? Was the egg always there to produce the chicken, or was the chicken always there to produce the egg? One of them had to be there first without a first cause. Likewise, at some point we must be willing to accept that nothing could cause the first cause. Can we get something from nothing? Probably not. But can we get something from a first cause? Definitely! Whether we want to define that first cause as energy, matter, God, or anything else is irrelevant at this point. For the moment, all that needs to be grasped is that a first cause is a necessity for anything involving a cause-and-effect relationship, which of course we experience on an everyday basis, in the lab as well as out in the universe.

By the way, most physicists today view the loop perspective as the correct solution for the paradoxical question (specifically for the reasons of parsimony, cause and effect, and the conservation of mass and energy).[37] This is one of those areas where I may seem to contradict myself, but if you understand how we arrived at this place, then you'll also see how the contradiction fits within the whole puzzle and is no contradiction at all.[38]

---

37 If we assume a preexisting universe prior to the BB, based on what we already know about our own universe, we would deduce that the preexisting universe was in a state of high entropy. In actuality, this is not the case. Our closed system universe started out as homogenous and a finely tuned system (that is, in a low-entropy state). The knowledge that our universe started from a low-entropy state goes against most of the physical laws as we know them. Can we attribute this beginning to chance? Maybe, and maybe not. From a strictly logical, cause-and-effect perspective, chance occurrence is unlikely. And even when seemingly chance things do occur, it's usually because certain variables have been controlled for. In other words, for a chance occurrence of an ordered state being created from a disordered one, the variables of matter, energy, momentum, and space would have to be controlled in such a way that a chance burst (a BB) could happen (as opposed to a noncontrolled state in which a chance burst is next to impossible). Also, who's to say that there was a preexisting universe prior to ours (which is an assumption)? We're pretty sure something was preexisting, but to conclude it was a universe like our own (whether in a high- or low-entropy state) is a bit presumptuous.

38 Please refer back to the preface concerning contradictions in explaining things from a nonlinear perspective to a linear thinking world.

The Not-So-Obvious Matrix

So far we have been discussing the physical realities of the Matrix, but those physical realities are only the tip of the iceberg. Just as physicists use experiments to observe cause-and-effect relationships, social scientists use the same methodologies in trying to understand what it means to be human. Of particular interest to our discussions will be those cause-and-effect relationships that specifically deal with human reality and our thoughts and actions within that reality.

We now know that the connections that tie humans to their environment (their actions, observations, and ultimately behaviors), are all interrelated. Just as we do not fully understand the interconnectedness of space, we still do not fully understand the connections that constrain and support us within the Matrix. What we do know, however, is that these connections are universal in nature, unseen, and associated with very deterministic qualities.[39] As a society, we have not fully grasped the concept that everything—object, action, and even thought—is interconnected and interlinked with a reality we cannot see. Absolute truth is and always has been a stand-alone reality that is independent from our relative definitions. Were it not so, we would have absolutely nothing on which to base our relative definitions. It's not only the absolute reality that existed before the BB (long before humans appeared), but also the reality that currently gives everything life.[40] Jean-Paul Sartre once observed that "no finite point has any meaning unless it has an infinite reference point." And this is exactly the point from which the Matrix emanates, the infinite reference point,

---

39 One example among many: Dr. Martha Farah of the University of Pennsylvania, and Dr. Michelle Schamberg and Dr. Gary Evans of Cornell University, have confirmed that children born into poverty (living in chronic high-stress conditions) grow up with less developed (even damaged) brains, compared with children raised in middle-class families. Their studies demonstrate that, connections with the environment, with other people, and with previous decisions and actions all have a bearing upon the Matrix and absolute reality.

40 This is a deduced truth that is beyond our concepts of empiricism and idealism (i.e., Enlightenment methodologies).

an absolute reference point of truth that always finds its origin within the infinite.[41] It's not unlike the link between general relativity and quantum mechanics, whereas the totality of the Matrix is almost beyond comprehension. It's a place where life and death, time and space, past, present, and future are all woven together in ways that no one fully comprehends. Absolutely nothing is done in one part of the Matrix that does not affect another part. Everything about us, our world, the universe, and our existence depends on and is connected to something within the Matrix: nothing stands alone. To value or devalue one thing within the Matrix is to ignore the interdependence and interconnectedness of all things. In short, the Matrix conveys an immensity and complexity beyond our usual awareness, much in the same way as our unconscious mind is beyond our conscious self. As you will see in our continuing discussions, the human interconnectedness within the Matrix is of significant value in understanding ultimate truth. We could spend vast amounts of time discussing the Matrix in scientific terms, but it would be a grave misuse of our time. Consequently, we are going to focus directly on how humans and the Matrix interact and how that interaction has affected our lives and the continual destruction of our planet.

From a physical standpoint, the most challenging problem humanity faces today is the population explosion. In the not-too-distant past, with few exceptions, humans did not concern themselves too much with the interconnectedness of the Matrix. At that time, the human population was dispersed among isolated communities and societies. If for some reason the structure (the Matrix) collapsed in one society, its surviving members could either start over again (somewhere else), or they could simply stay put and succumb to the gradual unraveling of the Matrix around them (that is, environmental, social, and political collapse). Now, due primarily to the

---

41 If this statement seems contradictory, it's because it is expressed linearly so as a linear mind can better comprehend it. Remember, expressing something of a quantum nature to a linear-thinking mind will invariably lead to many a supposed contradiction where none truly exist.

population explosion, we no longer have the option to move to a place with a viable Matrix structure that is still healthy enough to support us. Earth's population has grown to the extent that the Matrix is being affected universally much more than ever before. For example, carbon emissions have passed from the point of regional concerns to that of global distress. The world's oceans have been overfished and polluted to the point where it affects all of humanity and not just a few isolated populations. Whether we like it or not, we have become one global community. Hence, understanding the ramifications of our interconnectedness goes way beyond mere local, political, or philosophical speculations. In our current state of affairs, understanding how the Matrix works and then implementing a plan to save it will ultimately mean the difference between life and death. To highlight this point, let's spend a minute talking about global bifurcation.

The branch of applied mathematics known as bifurcation theory has been used in economics to predict such things as stock market crashes, in the medical field to predict when an epileptic seizure will take place, and in environmental biology to predict catastrophic changes within ecosystems. The idea behind bifurcation models is that just prior to a major change or collapse, a system will experience extreme increases in variances (e.g., high variability, odd movements, erratic behaviors). Whether the increased variances are extreme fluctuations within an ecosystem, radical movements of the Dow Jones Industrial Average, or increased oscillations in brain wave activity, these variances are all *measurable* erratic events that occur prior to catastrophic changes. Accordingly, bifurcation modeling gives the researcher unmistakable warnings, an early signal of impending breakdown within a system.

Global bifurcation is what the world is currently experiencing, as demonstrated by such things as the increased frequency and intensity of storms, droughts, earthquakes, dead zones on land and ocean, global warming, glacial and continental ice meltdowns, pollution, and depletion of water supplies. All of these are the early warning signs of an impending ecosystem collapse; the only difference is that the ecosystem affected is not small or isolated. It's nothing less than the ecosystem of the entire planet. Humans have destabilized the natural order, health, and interconnectedness of it all. Food webs have been broken or artificially changed. Erratic changes in temperature gradients, atmospheric structures, and ocean biology have led to what is quite possibly irreversible damage to the Matrix. These erratic changes are not static; they are cascading blows, a domino effect that will touch all forms of life on earth. These physical changes bring mental changes as well, as everything tries to adapt and cope with the extreme fluctuations that in turn cause changes in how we view ourselves, our neighbors, and the world at large.

If you've ever read *A Sand County Almanac* by Dr. Aldo Leopold, you cannot miss the Matrix connections between humanity and the world in which we live. Professor Leopold draws connections between human and animal, between human and land, between human and society, between humanity and self, and ultimately between life and death. I will not detail how Dr. Leopold provides the undeniable evidence of these connections (if you have not read his book, treat yourself to an extraordinary gift of wisdom), but I will say that he leaves no doubts as to humanity's influence upon the Matrix.

This brings to an end our discussion concerning the Matrix of Life. We could discuss the Matrix and its implications in a book all by itself, but that is not my purpose here. The key take-away to remember is that the Matrix is an absolute reality that's connected to everything, and gives life and order to all.

– TO TERMS WITH HUMAN REASONING –

I often debate the merits of showing others the whole iceberg and not just its tip. It is one thing to take an adventure upon ice-covered seas, quite another when you're in the water, and entirely a matter of survival (whether mental or physical) when you go under the icy surface. I've noticed over the years that the people who are the quickest to dive in are also the ones who get chilled the fastest, surface the quickest, and thrash about the most. These are also the people who, once ashore, look at you as if you were mad for even suggesting that they take the dive in the first place. And, of course, from that moment on you gain a reputation for being insane—or worse.

Before going beyond this point in our discussions, you need to be aware of an incident that happened to a friend of mine many years ago. John wanted to dive deep, to see the iceberg from the bottom up, so to speak, so I obliged him and we began to have many a conversation that took us ever deeper.

I will not go into all the details of John's character, but suffice it to say he was one of the smartest people I've ever met. He graduated with a PhD from Harvard or Cambridge or some such institution where brains are nourished and fed all too well. I believe his master's was in applied mathematics and his PhD was in some kind of applied physics. John was one of those truly gifted people who had an excellent brain for logic and theory and an unmatched enthusiasm for mysteries. I was convinced that John was ready for the plunge and was eager to take him to the darkest and deepest depths.

John was the most accomplished diver and swimmer I'd ever met. He was a whiz at making the connections and drawing relevant conclusions. I was fascinated by the thrill John experienced with each new nook and cranny he discovered and explored. We surfaced from time to time, and I would always ask John if he wanted to go deeper and see more. His answer was always a resounding yes. As the days and months went by, we would dive deeper and deeper, until we reached those places where darkness prevailed over the light. I had told John before our last dive that we were going to a place where he would see nothing with his eyes, but everything with his mind, a place where the ocean currents met and demanded everything a person could give just to swim in place, let alone to move ahead. Furthermore, I explained that this place was one from which he would come back a changed person. The change would be about beginnings or endings, some good, and some bad. I gave him this warning actually believing that he would return with excitement and strive for new beginnings in his life. Unfortunately, that did not turn out to be the case.

When we finally returned to the surface, after what seemed like an eternity in the abyss, John was definitely different, and I could tell his mind was tormented. The truths presented to him in the depths of his mind were too much for him to handle. John was angry with me for taking him to a place that was unpleasant and hard, that required too much of him. After John left my care for something bigger and better, I heard about him from time to time. By degrees, John became a bitter and hollow man. He did everything in his power to expel the knowledge he gained from our dives. He ran from it, denounced it, and tried to forget it by indulging himself in comforts and pleasures wherever he could find them. But, like Dorian Grey, John held onto things he should have let go of in the abyss. To this day, I do not know what it was that happened to John in the dark regions of his mind. He never told me, and I never asked. Each person who dives into the abyss faces their own demons or angels. The abyss is deep and dark for a reason: each mind that travels to such depths always bends in some particular fashion.

Many years later, I heard that John went mad and was committed to an asylum in New York City. A few years after that, I heard he was released due to cutbacks in social services. Eventually John ended up at one of those institutions in which the mentally awry tend to gravitate: corporate America, politics, or the army. In John's case it was corporate America.

The very last thing I heard about John was by way of a friend, who ran into John at a conference in Houston, Texas. It turns out that John became the CEO of an energy trading company and was now swindling millions of dollars from ratepayers across the nation. I reflected on this for a few moments. John was now a highly paid and highly educated crook, who was committing larceny against an uninformed public with the full approval and endorsement of the American government. Indeed, he was one of the initial movers and shakers in the deregulation of the electric industry. I have to admit, this path was one of pure genius! He, along with other CEOs, lobbyists, and congressional leaders, created one of the greatest smokescreens the United States has ever seen—a cloud so thick that it completely hid the swindlers from the swindled, and baffled everyone except the creative geniuses pulling the levers. My sincere apologies to the American public for unfettering a mind that was best left to science.[42]

Going forward, we can discuss many of the hidden aspects of life, things learned in the depths, many of which can be used against humanity or for it. I've found over the years that in too many instances, the essence of knowledge is missed, misused, or rejected in favor of an easier path. That concern is always in the back of my mind. My wish of course is that the knowledge shared in our discussions will be used for good. Good in the way of opening one's eyes to a bigger picture, good in maturing us as individuals, and good in that the truths of life will be of practical use to us and not misused by others.

The force of modern orthodoxy is against the person who demands that the truth be heard and applied. But in order for us to use and understand that truth, we must be willing to enter the depths and face the demons that hold us back as individuals, as nations, and as a planet. You are about to take that first step into some very dark and deep places. You will be tried in more ways than one, and once there it will be tempting to turn and run. You may feel regret, anger, depression, or any of the other

---

42 On a side note, there are some businesses, CEOs, and government officials that do the right thing when it comes to truth. Three in particular come to mind: U.S. Ambassador Joseph Wilson, Brooksley Born, Chair of the Commodity Futures Trading Commission (CFTC), and Ray Anderson, America's greenest businessman. If you've never heard of these people, there's a good reason: we live in a society that promotes the status quo, even to the point of suppressing the truth.

emotions imparted to our being. Remember that a big part of traveling to the depths is being able to confront your emotions in a way that makes a place for wisdom to grow. If you can pass through this darkness, you'll arrive on the other side knowing the true meaning of many things.

So, why are we entering these dark depths, and why is it dark in the first place? I'm glad you asked! The basic answer to both questions is that part of the journey deals with you and me on a very personal level. It deals with parts of our personalities that rarely if ever see the light of day—parts that have remained hidden, and hence there's a darkness to those hidden places. These realities are often unknown to us, and oftentimes they're next to impossible to detect. That being the case, we are not diving into the depths simply to acquire more knowledge; quite to the contrary, the truths we will uncover are those that have the biggest impact on our lives. Sometimes we may see a shadow of these hidden truths, but we seldom get the chance to see those things that control us in their entirety, in ways we never thought possible. And in order to reach those depths, we are going to spend a fair amount of time exploring the world of our subconscious minds.[43]

---

43 Since we will be referring to correlation studies in future conversations, a comment or two should be made in reference to the social sciences and the drawing of correlational conclusions from correlational examples. While it is true that drawing cause-and-effect connections between two variables does not necessarily prove that one variable causes the other (as in correlational studies), it does not follow that a cause-and-effect relationship cannot be extracted from the initial correlation. Herein lies the nexus of many a missed opportunity in the social sciences. A correlation is merely a starting point, the trailhead by which one enters the forest before exiting on the other side. One can start with a correlation, then move to inductive methodologies (i.e., conclusions based upon experience or history), and finally end with deductive systems (i.e., reasoned outcomes where the effect[s] are necessitated by the cause). Hence, definitive conclusions can be made from correlational studies as long as they are followed up with inductive and deductive methodological systems of analysis.

- OUR SUBCONSCIOUS SELVES -

"I know 'tis but a Dream, yet feel more anguish
Than if 'twere Truth. It has been often so:
Must I die under it? Is no one near?
Will no one hear these stifled groans and wake me"

Samuel Taylor Coleridge

We often fail to consider the subconscious part of the human mind when we examine human perceptions, thoughts, and behaviors. The reasons for this are understandable. The subconscious mind is not well understood. Yet we know enough to know that it plays an immense role in the way we think and make decisions.

In most cases, our subconscious world is beyond the grasp of our intuitive reasoning abilities. Like an island in the sea, our conscious minds operate on the surface of the sea. But the island only happens to be the uppermost part of a mountain range protruding from the ocean depths.[44]

The subconscious mind is the place where our innermost desires and insecurities lie. It seems to easily reveal itself while we are under hypnosis or dreaming, but during waking hours it's that hidden reality which affects us in covert ways. For the most part, our conscious mind takes in outside information and then organizes it by way of our subconsciousness. In the course of the organizing process, the information passes through several filters, through which our subconscious mind adds internalized information to the new information, which in turn is used for decision-making purposes. The internalized information that we already possess is usually *emotionalized* information.[45] This is why we sometimes do things that are irrational and counterproductive. The subconscious mind blunts and

---

44 For an excellent description of how the subconscious mind remains hidden from us and how it influences our thoughts and actions, I'd recommend the book, Incognito, written by the neuroscientist, David Eagleman.

45 This emotionalized information is often repressed, denied, or forgotten on a conscious level, but is deeply rooted in the subconscious mind.

filters our perceptions of reality, which happens to be a very real factor of all thinking styles.[46]

Many behavioral studies and experiments have shown the control our subconscious minds have over conscious thought and decision-making abilities. Let's take a look at a few of them:

- The neuroscience community has long known that our subconscious world contains what's known as the "mirror system," a component activated by visual observations that in turn activates our emotions, subsequently inducing us to feel as if we were physically taking part in the observed situation even when we're not. The implications of this are astounding. First, if the visual cue touches upon those regions of the brain that give us pleasure (i.e., being competitive, taking risk, feeling excitement, imparting happiness), the images themselves can reinforce behaviors. Second, the mirror system explains why viewing socially good or bad behaviors can induce people to mimic such behaviors.[47]

- Asch, Milgram, Deutsch, Darley and other social scientists have shown that the subconscious mind controls a good deal of how we think and act. Elements such as social conformity, moral disengagement, unconditional obedience, informational influence, and the Genovese syndrome all attest to the power of the subconscious mind to overrule objective realities.

---

46 Thinking Styles to be discussed shortly.

47 Directly related to this is the role Activity-dependent plasticity plays in learning and behaving. Activity-dependent plasticity is the brain's ability to undergo physical change dependent on how the brain is used for different activities. Here's a question to ponder; if we are repeatedly exposed to certain events that focus on dysfunctional behaviors (i.e., lewdness, disrespect, promiscuous activity), will our brain undergo a change that eventually reinforces the bad behavior(s)? Current research suggests it does!

- The subconscious mind also plays host to and anchors quite a few of our cognitive biases. For instance, repetition bias is the willingness of the subconscious mind to believe and accept that which has been continually fed or repeated to it. In other words, if you are told something often enough, you will eventually factualize it regardless of any reality or truth that denies it. Attribution asymmetry is another major bias that controls our thoughts about ourselves. Our subconscious mind uses attribution asymmetry to attribute our weaknesses or failures to external factors and our successes to our innate genius or physical abilities.

In essence, the subconscious mind does blunt and filter our perceptions of reality in very real, overt, and covert ways. Even people who know how the subconscious mind exerts its control still find it extremely difficult to be objective when processing information and making decisions. The vast majority of people are simply unaware of the subtle influence of their subconscious worlds on their conscious reasoning abilities. Advertising, marketing, and propaganda exploit our subconscious worlds by manipulating a person's subconscious landscape. Images and sounds are the primary pathways for influencing and changing the ways we think. It's an established fact that the mere repetition of an idea, regardless of its truthfulness, eventually seeds itself within our subconscious mind.

This is a key point in understanding human psychology and humanistic reasoning. If a concept or idea is publicized, televised, put into music, or assigned to print, and repeated over and over again, then sooner or later the subconscious mind starts to accept the idea as reality based on the repetition alone.[48] The ideas move from concept to

---

48 This method of manipulating the mind is so common that one would think society would be aware of it and avoid its pitfalls. Political campaigns (particularly negative ones), military training, entertainment programs, marketing material, and other intentional influencers continue to use this technique to indoctrinate and condition the mind—and I might add, to great success!

certainty, and eventually this certainty leads to a pseudo reality. The effects of this conditioning are invisible to the person being changed or to the society that is being manipulated. I've witnessed over the years how mass media has exploited humanity's weaknesses, it has been a sad situation to watch. Let's take a moment to discuss how pervasive this type of conditioning is and how it affects what we view as truth and reality.

The Land of Multimedia

This puzzle piece seems the smallest, the most benign, but it has proved to be the biggest and deadliest of them all. Many times we see what we do not see and are blind to that which stands boldly before us, and therein lies the danger. To many, this subject is pure folly. Even so, more insignificant things have become potent forces changing the course of world history. I'll begin with some personal history that has a direct bearing on what we'll be discussing below. In doing so, I hope to draw a more complete picture of where we're at today and where we'll surely be tomorrow.

I was privileged to witness the beginnings of television. My earliest memories of watching TV were on a black-and-white screen. I remember watching cartoons and Westerns in the later 1960s and early '70s, *Sesame Street* and the original *Star Trek*. In these early years, TV shows portrayed a fairly high value system. Comedy shows never focused on the ignoble or vulgar. Dramas were based on admirable outcomes, and viewers were left with positive reinforcements for doing good in the world.

By the end of the 1960s, a subtle but noticeable change took place within the TV and music industries. The Beatles epitomized this new change in attitude, a shift from being other-focused towards being self-focused (i.e., from a group focus to an individualized value system). The change was subtle, in part because much of the countercultural movement of the 1960s and '70s had much-needed altruistic goals in mind (e.g., equality issues, stopping corruption in government, protecting the environment, etc.). But these altruistic values masked a darkness.[49]

Growing up during this era, I embraced these countercultural changes wholeheartedly. There was no doubt that the government had major problems with corruption, the environment was being destroyed, and equal rights for African Americans and women needed to be ensured. Also, I enjoyed dressing the way I wanted to, growing my hair long, having the freedom to explore sexual liberation, and smoking marijuana with friends. Overall, I had a great time during the 70's.

However, in the later 70s and early 1980s, the darker side of the countercultural movement started to reveal itself in new ways. The things that once gave society civility and respectability began unraveling. Individual rights started replacing societal rights. Recreational drugs began creating generations of users, spurring crime, violence, and deaths that continue to this day. Billions of dollars have been spent and lost as a direct result of the acceptance of those dysfunctional attitudes and behaviors. Prior to the 1950s, the darkness we deal with today was practically a non-issues (e.g., single-parent homes, latchkey children, escalating crime rates, increasing incidence and new varieties of sexually transmitted disease, and a rise in the use of illicit drugs).[50]

---

49 This smoke-and-mirrors phenomenon is used in quite a variety of ways. For example, a few parts in a TV show will always focus on taking the high road in life (i.e., using positive role models and doing what is morally good and noble for society). But these elevating values are no longer the norm in the media world today. The new norm is dysfunctional behaviors that promote tolerance for degeneracy, apathy, and cynicism. Because of the few noble qualities being introduced here and there, we tell ourselves that a particular media outlet (i.e., TV, music, sports, movies) is acceptable even while those darker elements infiltrate and condition our minds to accept ever-increasing levels of dysfunction. Misdirection is the fundamental tool of the magician to which mass media could be called the master.

50 It should be noted that prior to the 1950's there were problems dealing with teen pregnancies, alcoholism, rape, divorce, single-parent homes, etc. These instances however were far and few in-between as compared to what came after the 1960's. Crime rates soared exponentially after the 1960's, not because they were now being reported, but because social restraints that were once the norm were now removed.

The individual's right to free sex has taken such a human toll that it's impossible to count the number of individuals who have died or been permanently injured by so-called sexual freedom in society.[51] As of this writing, more than 25 million people have died of acquired immunodeficiency syndrome (AIDS). Recreational sex and recreational drug use have caused the deaths of billions and the broken lives of untold millions of people, all a direct legacy from the countercultural movement to having free sex.

Music and TV shows became progressively worse. Content was increasingly meant to shock, whether through foul language, the exploitation of depraved behavior for entertainment, the valorization of violence in images and lyrics, or other dysfunctions that drew audiences to talk shows, movies, music, and newspapers. Soon there were cartoons that glorified defiant behaviors in children and sitcoms that idealized the sarcastically rude, the sexually promiscuous, and the unabashedly self-centered.[52]

The media we unconditionally embrace has a power over the human psyche unmatched by anything in our world today. If you doubt this, perform this simple test to see just how much control the various forms of

---

51 Free sex is having sex outside of the bonds of marriage between a man and a woman. This might be unpopular to state, but all the evidence points to its validity (i.e. cause and effect relationships, history, etc.). I'm not stating my opinion in this regard, as with all of our other discussions, I'm only stating what the evidence shows. People who remain faithful within a monogamist heterosexual relationship do not acquire sexual diseases.

52 Before we leave this section, it should to be noted that when dysfunctional thinking and behavior become the norm, one can expect as one of the consequences an increase in crime. This is exactly what happened as the fruit of the counterculture movement began ripening in the late 1970s, through the 1980s, and into the 1990s. Crime rates exceeded anything experienced in approximately the first two hundred years of United States' existence). Many social scientists associated the rising crime rates with the implementation of tougher laws and the increased number of convictions that followed. What these social scientists rarely asked, if ever, was why there was a desire to make the laws tougher in the first place. The actual reason for the increase in crime and the tougher laws was the Pandora's box that was opened by the countercultural movement. In other words, those things that have been known throughout human history as detrimental to individuals and society had been released upon an unsuspecting populace. The ideals of "freedom" were used as a smokescreen to cover the dysfunction unleashed upon humanity.

media have over your life. The test is simply this: go without viewing or interacting with any visual or auditory media for three months. This means giving up television of any kind (news, sports, dramas, sitcoms, cartoons, all of it), giving up video games, giving up going to the movies, giving up using the Internet, giving up listening to music. If this test sounds unreasonable, then you're more than likely under the control of the media devices and programs you own. It's a valid test that reveals who's really in control of whom: you the media, or the media you.

You may be surprised to hear it, but thousands of people in the Western world never use the media that besieges society today. The bottom line however, is that billions more are obsessed with the various forms of media that surround them. Media consumption becomes an addiction that works slowly, methodically, almost imperceptibly to control the mind. As with most things created by humans, the design did not begin in a corrupt state; media was created with good intentions. Radio shows, TV shows, and music began for the good of society (much as the countercultural movement began its life).[53]

Prior to the 60's, mass media kept to certain moral values. Once again, just view or listen to the TV or radio programs of those times. They had a form of innocent purity that is completely missing today. Today society has an insatiable appetite for anything new. Retailers know this fact well, as do the marketing departments at TV networks, music companies, websites, and anywhere else that sells a product. It's an established fact that humanity's greatest strength and ultimate weakness is our desire to seek new knowledge and new experiences, and to follow the latest developments and news despite its sources or agendas. This desire for the new is fanned by today's mass media and, increasingly, by the devices we use to keep us continuously connected to that media. After all, that's where the

---

53 While it is true that TV and radio began with the ideas of making profit, it is also true that they began with the ideals that were already established in society, that of seeking a higher, nobler position for humanity, that is, having and keeping a healthy family life, serving others, having high values, etc. If one seeks evidence for these types of values, they can be found in the content of TV and radio shows from that era.

real money is to be made, and that, my friend, is our Achilles' heel and in all likelihood our eventual undoing. Does this necessarily mean that all modes of multimedia are bad? Certainly not. Unfortunately we don't just view the good; we want to explore and experience everything, the bad as well, regardless of its effects upon our mind.

Gaining Access to the Psyche

From an early age, various media forms indoctrinate us, creating viewing habits over which we have little control, thus the socialization process for the child begins.[54] If our parents do not guide our viewing habits, or are incapable of doing so, we internalize the values (or non-values) on an individual basis, and those values or lack thereof eventually spreads to society as a whole.[55]

As children and teenagers, we long to fit in while still being noticed as individuals. To do this, we watch, learn, and then act accordingly. Most of this watching and learning takes place through various types of visual media. Sitcoms, reality shows, movies, music videos, and the like, which typically portray a world and lifestyle that is focused on being fashionable, attractive, cool, rich, and powerful, the focus is on self-gratification. Rewind to the days before 1960, when most families remained intact bastions for value training. The goal of parents was not to stress material success, vulgar language, and the pursuit of fame and fortune. Quite the contrary, their goal was to instill respect, humility, and a drive for magnanimity. So why have we been so deceived by the popular media of our day? There are two key drivers behind the change in the value systems of past and present generations. The first is

---

54 This all happens at an age when we do not yet have the mental capacity to imagine the eventual consequences of the habit.

55 There is a great disconnect in our present age from the parenting practices of pre-countercultural generations. TV and other forms of media tend to be used as electronic babysitters for children, leaving children to be taught by media sources that continually extol self-serving attitudes and behaviors.

based upon Enlightenment doctrines, and the second is based upon self-gratification dynamics.

The Enlightenment

The Enlightenment is usually defined as a cultural or philosophic movement that took place between the seventeenth and eighteenth centuries. However, it's good to remember that Enlightenment thinking and methodologies are still with us today; in many respects, they never left. Enlightenment thinkers wished to reform institutions, such as the church and the monarchy, that were based on principles of tradition and faith. A reformed society was the goal of enlightened thinkers and planners. In many ways, the Enlightenment was the catalyst for cultural change within westernized societies as its ideals and methodologies spread from the philosophies of intellectuals, to educational institutions, to governments, to the sciences, and finally as a consequence, to the populace as a whole.

The Enlightenment based its belief system on individual experience in an effort to define truth and reality. If an event or a concept could not be tested by observation or experimentation, then its claims to represent knowledge and truth were suspect. The word *empiricism* is derived from the word *experience*; hence, to experience an event with one's senses was to encounter it empirically.

Empiricism eventually supplanted rationalism, which emphasized discovery by way of reason alone as a way to determine truth and reality. In other words, the evidence of the senses was preferable to the evidence deduced by the mind alone.

Thus, the Enlightenment produced a colossal change in the way today's society thinks, acts, and views truth and reality. The old ideas of truth being discerned a priori (before the observation or experience) by deductive processes, was replaced with the new ideal of truth discovered a posteriori (after observation or experience). Because of this transformation in thought, any individual or institution that based reality on rationalistic concepts was scorned, ridiculed, and in some cases destroyed in favor of the new orthodoxy, namely Enlightenment ideals.

The American and the French Revolutions were directly tied to those ideals. Monarchies and religious orders that claimed divine right to rule (a form of a priori supposition) were now deemed irrelevant and even dangerous to an individual's freedom to discover truth for themselves. Thomas Jefferson, Benjamin Franklin, and many other great thinkers were great proponents of Enlightenment ideals. Much of the U.S. Declaration of Independence and Bill of Rights is the end result of Enlightenment philosophies and activism. Social contract theory, self-determination, secular morality, and social progressivism were all products of the Enlightenment and established as the core values for a modern age.

Little did the thinkers of the time realize the long-term corrosive nature of what they had unleashed. Knowledge gained by way of rationalistic methods suffered, and by putting the emphasis on experiential truths, the individual became the final judge of what was and was not reality or truth. Individualism slowly became the new moral code of life, each person deciding what is right or wrong in his or her own eyes. From a basic scientific standpoint, this ideal is good in the sense that many eyes (peer reviews) are used to verify experimental results. However, from a strictly societal standpoint, it becomes extremely complicated. As a direct result of the Enlightenment, the individual becomes his or her own judge and jury of how to live freely. Given our nature for self-gratification and self-meaning, the idea that we can judge what is and is not truth should be called into question. Is reality or truth dependent on recognition by a human, or is it independent of our recognition? If we follow Enlightenment principles to their ultimate conclusion, we find a reality based entirely upon perceived experience, in scientific inquiry as well as in individual life.

In the Enlightenment model, humans are the center. We put ourselves into the position of judging what is right, either individually or corporately. Hence, truth becomes dependent on the individual; the individual does not become dependent on the truth. This final empirical outcome is nothing less than the complete antithesis of rationalism. Once again, we enter a pre-Copernican state in which everything revolves around us as the anthropomorphic center of the universe. Surely Epicurus and Berkeley would approve. Rationalism soon becomes the stepchild to empirical beliefs, and truth becomes superfluous as each person defines reality in his or her own way.

The Enlightenment did achieve some good things, liberating humankind from the tyranny of self-imposed rulers and rules designed to keep people enslaved to a system (be it church, realm, or ruler). Freedom was and still is the rallying cry of the Enlightenment: to be free is to be free in all ways. Insofar as achieving a system in which total freedom reigns, we are basically there. Consequently, what we now learn via the education system, pop culture, and the media is that we live in a land of mirrors where we constantly watch ourselves.[56] The problem with this vantage point is that freedom reflects nothing but its own image—I am reflected so therefore I am. In this land of mirrors, there is nothing but circular reference points. So that, beyond the image nothing really matters. Discoveries that could be made beyond the mirrors are lost to the images reflected therein.

Hence, from a philosophical and psychological perspective, Enlightenment thinking always emphasizes what humans want or judge to be important. The core of this reality is exclusively based on experience. Unless an individual can experience something, it's either inconsequential or nonexistent. This type of thinking has a great deal to do with how we view ourselves and the world we inhabit. So, on a subconscious level, those who have been raised on multimedia and Enlightenment thinking automatically assess everything from the vantage point of self. Because of this self-centered relationship to reality, everything is extrapolated from what we sense. If I don't experience something with my senses—that is, if it's not all about me—then it's obviously unimportant. This is a huge

---

56 This does not necessarily mean watching ourselves as in being a narcissist, but as in watching ourselves in how we create and define things that prove we are the center of all things free and good.

stumbling block to our understanding of universal truths and objectified reality.

Self-Gratification Dynamics

From a marketing standpoint, the primary goal is to create a desire to please one's self. This is done by accessing an individual's subconscious mind, through which it is much easier to condition that individual to the idea that his or her desires, feelings, and happiness, are of the utmost importance. Although most people do not realize it, this conditioned response hides in the background of their subconscious thinking. Media sources hope to inspire the subconscious override of the conscious mind to elicit certain behaviors. These behaviors manifest themselves in many ways (e.g., in choosing a particular show to watch, listening to a certain style of music, purchasing a particular brand). When the media is geared toward capturing a person's interest and money, it always appeals to self-gratification. Like Enlightenment systems, marketing structures condition us to the idea that everything revolves around us as individuals. Hence, between Enlightenment ways of thinking and the media's intent to promote and sell, each philosophy or system reinforces the other, furthering our narcissistic belief that we are the foundation of all things.

Children and teens who grow up consuming the various forms of media are continually conditioned to entertain self-gratifying thoughts. This continual reinforcement shapes our thinking in a way that is unhealthy for the individual and destructive for society as a whole. The mentality of a person conditioned to continual self-gratification is that of a child, even while being in the body of an adult. Today, we have at least three generations of adult children within the United States, all thinking and acting in self-focused ways. To highlight this fact, let's look at some examples.

In the 1940s, 1950s, and early 1960s, if you were to attend a football game, you would notice that the spectators were both courteous and respectful in their behavior towards each other. At today's football games, you will hear foul language, you'd see many a drunken spectator, and you'd see very little respect for anybody; in fact, you might observe a drunken brawl between fans of different teams. None of this was a part of the spectators' experience in the early decades of team sports.

From the mid to late 1960s on, men and women started gravitating away from responsible behaviors (i.e. being considerate of others) and moved toward self-indulgent behaviors geared to egotistical ends. Over the last thirty-some years, I've witnessed more and more people spending money and time on the superfluous (i.e., luxury cars, motorcycles, large homes, video games) than on mature investments (i.e., saving money for retirement, spending money and time on those in need, etc.). So instead of learning how to be other-focused, self-effacing, and respectful of all people, qualities that truly give meaning and purpose to life, we value a society that is fashionably funny (via sarcasm and cynicism), constantly in pursuit of the latest fad or gadget, and eating themselves into super-sized giants.[57]

Mass media goes a long way in creating a singular consciousness. TV, music, movies, sports, and other media forms perpetually condition us to act in certain ways, appealing to our emotions and manipulating our attitudes, spending habits, and use of time. In essence, mass media controls the way we think. We don't like to admit it, but we let someone else do much our thinking for us, on both the subconscious and conscious levels. For the majority of people, that system is a way of life. Conversations revolve around the latest movies, sporting events, money, and prominent people. Discussions about the most important issues of our time are virtually unheard of.

---

57 Entertainment up to and including most of the 1960s was considered a luxury, something that happened occasionally, and only after more important matters had been addressed. Basically, all forms of entertainment were considered inconsequential parts of life that added little value to what was considered important (e.g. spending quality and quantity time with a spouse and children, cultivating the mind, writing to friends and family, reading to children as they drifted off to sleep, etc.).

Let's face the facts. It's just not very much fun feeding a baby born with AIDS, helping a street person get a new pair of shoes, fighting to abolish the sex trade, saving a child from poverty, or even reading a good book to a child while our favorite program is on TV. And isn't a free society allowed to be self-indulgent and self-expressive, even to the exclusion of 80 percent of humanity?[58] Freedom, without the responsibility that goes with it, destroys the beautiful things in life. If we can support the multibillion-dollar video-game industry or spend our hard-earned money on the latest gadgets, then why not? It's all mine anyway, right? I deserve these things because I live in a wealthy and free society and have the constitutional right to self-expression and happiness.

These attitudes infect much of U.S. society today and block our access to truth and reality. Practically all of them are the direct result of our seeming inability to remove those things that reinforce the "me" mentality. Our views on respect, humility, justice, love, pride, and freedom are quite different than they were just fifty years ago. There's a very old truth that is conspicuously missing in our current mental environment.[59] It's the simple moral ethic of reaping what we sow (cause and effect). Honestly, can we really watch the degradation popularized by today's media and presume to say that our values, ethics, and morals are on a progressive path towards improvement? Can we honestly say that we've matured as a people to ideals of living simply rather than in excess, and of prioritizing the good of others (i.e., justice for all) over the exclusive good of the few?[60]

---

58 A type of Pareto efficiency (see glossary for full definition), in which self-serving behaviors begin to make someone else's life worse.

59 Through the millennia, people like Buddha, Confucius, Jesus, and Gandhi taught the profound significance of striving for values that dignified and matured humanity for the betterment of every person and the environment we depend on. Much of what we reap as a society today is what has been sown by the media programming of the last half century.

60 Media's ability to waste vast amounts of time, time once used for the betterment of self, family, friends, and community, has diverted the lives of millions into a virtual wasteland of irrelevance.

The contrast between what came before the countercultural movement and what came after is nothing less than epic. Between Enlightenment views of ourselves and the media hegemony in our lives, we have all been deceived in one way or the other. Somewhere along the way we became lost. Where once we were on the high ground, now we tread a path that descends into darkness. Indeed, the path we now travel is more evil than any of us fully comprehend! Our fascination with and unconditional acceptance of everything that the media offers is literally blinding our vision. It is wisdom's role to recognize the illusion, the danger, and the necessity of change. Do we continue to do nothing and live in pseudo peace for a time, or do we take the hard path on which we will be tested, sometimes to our limits? Seeking truth is the easy part, but living it is where truth comes alive.

- THE DARKNESS WITHIN -

"The spirit who bideth by himself
In the land of mist and snow,
He loved the bird that loved the man
Who shot him with his bow."

Samuel Taylor Coleridge

Throughout history, intellectuals, politicos, and mobs have tried to legitimize their revolutions by accusing individuals, groups, or organizations of wrongdoings: the Jew by the Nazi, the capitalist by the socialist, the conservative by the liberal, the Christian by the Muslim, the religious by the secularist, the bourgeois by the proletarian, the communist by the capitalist, the royalist by the revolutionary, and on and on. The trouble with this worldview is that it always assumes the problem lies with someone else or is caused by some outside force. It places the insurgent, the rioter, and the avant-garde revolutionary on a course of growing arrogance, pride, and contempt that in turn blinds them to their own sins.[61] Why do you suppose that out of all the revolutions in the history of the world, not one of them has eliminated injustice or inequality anywhere, including in societies claiming to be democracies? If nothing else, these social movements and revolutions should point us to the fact that humanity's problems are not external in origin. Indeed, the deeper issue, the one that is too dark for most to accept, let alone examine, is that the majority of society's problems are internal to each of us as individuals. The reason that we never seem to climb out of our perpetual hole is that injustice, greed, rebellion, envy, pride, and inequality (all basic forms of narcissism) quite naturally lives within each of us. Some individuals may seem to live without such vices, but rest assured, given

---

61 In light of history, these particular sins have proved to be just as problematic as those railed against by revolutionaries.

the right circumstances, these dark evils will manifest themselves in some way, shape, or form in anyone.[62]

Everyone bears a degree of culpability for the situation in which we find ourselves today. The intrinsic problems we face stare stoically at us from our own mirrors. For it's how we think and act as individuals, as opposed to how groups or governments think and act, that is the quintessential problem for humanity. You can't change how you behave without changing how you think. In essence, each of us carries a liability, a burden that is easier forgotten than dealt with, easier ignored than highlighted, easier to see in others than in ourselves, and easier repressed within our subconscious minds than dealt with in the conscious light of day. And unfortunately, it's the root cause of most of the world's problems today.

The much deeper and more poignant realities that underlie the human condition are insecurity, duplicity, ignorance, hypocrisy, manipulation, fraud, fear, lust, envy, ego, retribution, and pride. Our conscious mind ignores or denies objective reality, making it extremely hard for the average person to see, let alone accept, that these elements are as much a part of us as the water we drink or the air we breathe. The subconscious mind conceals these parts quite well; they are nevertheless very real parts of who we are and substantially influence how we think and act.

---

62 Case in point: if someone were to call you a liar, you'd have to admit that you are indeed a liar. If that same person said you treated others inequitably, again you would have to unconditionally agree that you treat some people better than others. I'm sure there are those who completely disagree with admitting to such deficiencies. But if you are honest with yourself, you'd see the truth of the matter. We lie to spouses, coworkers, friends, bosses, and everyone else—white lies and blatant lies. The lies we're most blind to are the ones we tell ourselves (i.e., I have a right to do as I wish). Now let's touch on the second statement. Worldwide (as of this writing), approximately $60 billion is spent per year on video games. At the same time, approximately 6,000 children die daily because they don't have access to safe drinking water. Would you be willing to admit that there's some inequality going on when one person can spend $50 on a video game while another collapses in the dust and dies for lack of drinking water? The examples are literally endless. We all have some darkness that lives within us.

In many respects, our desire to distance ourselves from the darkness within is an admirable goal. But if those inner realities are never accepted as being part of ourselves, we will always be forced to wear masks that help us live in a self-created reality/non-reality. Authenticity—being a real person who accepts his or her inadequacies, those of others, and who lives at peace with the knowledge of being less ethical, moral, honest, altruistic, or loving than he or she wishes to be—is where absolute reality starts and true knowledge begins. If the starting point is anywhere else, our understanding of meaning, truth, and reality becomes biased and, in the end, extremely distorted.

Before we leave this subject completely, one more point needs to be made. In order for you to understand the truth and reality of anything, you must first acknowledge your own self-centeredness and ignorance.[63] Part of our inheritance as human beings is a great deal of foolishness, fraud, and arrogance. Hence, in order for me to even understand the foolishness of others (i.e., society), I must admit to having it myself and experiencing it firsthand, day in and day out. One of my goals in writing this book is for you to come to the realization that we are all weak individuals. Once you can admit the reality of the human condition, then and only then can you take that first step in understanding the greater reality that lies beyond. As you'll discover, a humble person is the only one who can mature and grow in truth.[64]

Collectively, we have the power to deny ourselves in order to make other lives less grim and more hopeful. Can we create such a world of beauty that transcends the corruptness that lives within us? We have the power to do it, but will we?

---

63 If you honestly do not think you are self-centered and ignorant, then you have missed the nexus of everything we have discussed to this point. But for those who understand the logic and reasoning leading to the conclusions we've made thus far, the remainder of this book offers some profound insights for you.

64 A humble person in this case is probably the very person who recognizes this reality in the first place.

Indeed, everyone must address and accept those things hidden within our lives before any advancements can be made. The evidence will be plain enough as our discussions progress, but its acceptance will rock most of you to the very core of your being. Why? Because our minds do not want to accept that which is painful to behold or the implications that spring from such visions. We can walk into the darkness of our minds to face those truths that bind us to a false reality, or we can remain imprisoned forever, never to know the full meaning of life or to achieve our totality in it. Come now, and let's enter those hidden places and spaces together, and let's endeavor to free our minds of the chains that imprison us to the subjective reality of self.

– THE SIX THINKING STYLES –

Cognitive psychologists, behavioral economists, and sociologists know that people are irrational creatures who systematically fail to comprehend the outside forces that guide their thoughts and actions (one of those forces being the subconscious mind). This includes all of us, from the most academic to the least educated. As much as we would like to believe otherwise, it is a proven fact that most of humanity operates from a self-focused perspective (i.e., I decide what is truth or fiction), while disregarding objective truths that transcend each individual. As we'll soon discover, this kind of focus is reasonable in light of our history and current standards. But however that may be, this seemingly reasonable line of reasoning ends up obstructing our ability to see reality for what it is.

People have subconscious preferences, and they will tend to use preferred methods of reasoning on a consistent and regular basis. These preferred styles of reasoning anchor our thoughts and actions. It is not an absolute way of thinking, nor is it an unchangeable way of thinking, it just happens to be a preferred method of thinking.

On a very basic level, scientists have discovered that humanity can be divided into two general thinking subgroups, the first group being the abstract thinkers and the second group being the concrete thinkers. Abstract thinkers look for concepts or ideas, while concrete thinkers process the observable facts before them. That is, abstract thinkers think outside the "box" of a created system, in this case the limitations of what society largely believes to be true and real, i.e. a Newtonian or pragmatic perspective on life.

Overall, concrete thinkers are more concerned with questions of "how." By focusing on the how, the concrete thinker can use reductionist methods (simplifying and reducing outside variables) to gain clarity. The irony is that this type of thinking almost always leads the concrete thinker down a path of self-assurance and, in the final analysis, to a pernicious form of blind faith that seemingly has been proved, and is therefore irrefutable.[65] Concrete thinkers have a very hard time answering "why" questions, which always deal with complexity that cannot and should not be reduced or simplified in the quest for answers.

But the complexity of the human condition extends far beyond the simple concepts of these two broad categories of thinkers (themselves part of a concrete methodology). In order to truly understand humanity, we must drill further down until we reach the core issues, not only of the hows and whys, but also the interconnected layers of human reasoning that go beyond these simple questions.

With that in mind, the boundaries I've drawn below between the various thinking styles are by no means static. My goal is to highlight the major differences in thinking methods. In reality, the various thinking styles do not have definitive boundaries. People are dynamic beings who can have moments of insight, who can process information using thinking styles other than their preferred style, and who have the ability to permanently change from one thinking style to another or combine several styles into one.

Individuals process information in different ways, depending on a host of environmental variables, as well as the biological ones we are born with. But we can still draw generalized conclusions about preferred ways of thinking from observation of the population as a whole and characterize society by the way that we corporately process and apply information on an individual level. So, with that said, let's turn to the six thinking styles of humanity.

Style I - In this style of thinking, people tend to defy and disobey authority, not out of reason, but for narcissistic ends. Detailed thinking beyond self-interest is practically nonexistent. Very young children, a significant

---

65 As our journey progresses, you will soon realize that a proof can be correct in one domain and completely false or irrelevant in others. The notion of a so-called proven concept must be weighed against all the complexities and interconnections that make up the universal whole.

proportion of the criminal population, and many people who suffer from mental illness fall within this category. Style I thinkers have a very local awareness of their surroundings. This includes knowing where they live, what they do for a living, what day of the week it is, and other immediate concerns.

Style I thinkers make up between 10 and 15 percent of the population (mostly children).

Style II - In this style of thinking, a person trusts in ideals based on social norms or traditions established by accepted authorities. When a person learns to obey authority (i.e., parents, law enforcement officials, employ-ers), they have moved from Style I thinking into Style II thinking. Ideals put forth by political, religious, and philosophical movements, or any ideal that is fully accepted and embraced without thorough knowledge, fit with-in the parameters of Style II thinkers. Some detailed thinking does occur within this style, but is usually limited to efforts to preserve the values of the community or personal philosophy of life. These thinkers therefore lack the intuitive ability to understand the more complex questions of life or anything outside of their own perspectives. Are Style II thinkers stu-pid? Absolutely not. There's a great gulf between stupidity and lack of awareness. I've known mathematicians who were geniuses in their chosen field, but when it came to most other things in life, they were still Style II thinkers.

The overriding characteristic of Style II thinkers is the strong desire to be accepted and approved by others.

Style III - This style of thinking involves broader perception levels than Styles I and II. This includes a degree of knowledge of the nuances in-volved in social relationships and cause-and-effect relationships. However, these thinkers are still limited in their ability to synthesize across large amounts of information. By this I mean holistically looking at the broad picture that unifies life. Style III thinkers can and do think deeply about certain things, but they still miss many of the far-ranging interconnections that make absolute truth a reality. Style III thinkers retain many of the ideals of Style II thinking, but they temper those ideals with a greater base of knowledge (not necessarily a broad spectrum, but a narrower, more concise knowledge focus). This thinking style may retain doubts about specific ideals, but they're still unwilling or unable to express those doubts in a public forum due to their strong desire to be accepted by their peers and community. Style II and III thinkers always lean toward conforming with an accepted system. For the most part, these people are "comfort conformists," comfortable with their established ideals or systems regardless of the truth (or lack thereof) that under-girds those systems. These individuals desire harmony and order, and will compromise truth to maintain order within those accepted systems. In most cases, Style III thinkers will ignore or reject difficult truths in order to maintain the status quo. Rules, values, ethics, and norms are defined by their chosen community, rather than by independent thinking. Because of this, Style III thinkers are particularly susceptible to groupthink. Style II and III thinkers are defined more by what they refuse to accept than by what they are willing to consider. Style III thinkers can be very smart and precise in many things, but they lack a global or all-purpose perspective when examining the foundations of their belief systems.

Style II and III thinkers make up the vast majority of society, and represent on average 65 to 70 percent of the general population.

Style IV - For these thinkers, objective reality, showing love and respect to others, and humility in heart and mind are innate qualities. Others' well-being is their primary motivating factor.

Style IV thinkers never approach a situation from a political perspective (i.e., What can I gain from the situation?), but rather from a position of neutrality (i.e., What yields the best outcome, regardless of personal preference?). Style IV thinkers might disagree with another's way of thinking, but they will maintain close and loving relationships regardless of others' thoughts or actions. The big difference between this thinking style and Styles II and III is the unwillingness of the Style IV thinker to turn a blind eye towards half-truths or dysfunctional behaviors. Style IV thinkers hold others accountable for their thoughts and actions, whereas Style II or III thinkers do not. The reality of this kind of thinking and love for others runs deep in Style IV people. Some of the most admired and respected people on the planet have been Style IV thinkers.[66]

Style IV thinkers have a very good grasp of the big picture; they discern a multitude of variables and then apply them to living a harmonized and fulfilled life. Because of this, Style IV people almost instantly and intuitively know the outcome of any given societal or personal situation. For the Style IV person, truth is tested and verified by people (i.e., truth is fully discovered by the consequences of a person's choice). If a choice leads to a less-than-healthy outcome, then that choice is deemed fallacious by the Style IV thinker. In the mind of a Style IV thinker, it is better to hurt someone in the beginning than to watch them ruin themselves in the end. Style IV thinkers will always tell you the truth and will always love you no matter where you are in this journey called life.

If there is a down side to Style IV thinking, it is that many people don't understand what motivates the actions of a Style IV thinker. This thinking style comprises about 1 percent of the population.

Style V - This style of thinking is used primarily by scientific and academic thinkers. They are independent thinkers who like to go beyond what is held as common knowledge. In Style V thinking, the person does not implicitly trust ideals or concepts without being able to systematically test their viability or truth. Even so, these thinkers usually maintain a modicum of Style II and III ideals of fitting in and not appearing odd or radical

---

66 Some of the more famous Style IV thinkers are: Buddha, Jesus, Gandhi, and Mother Teresa. Among the less well known are William Wilberforce, Dietrich Bonhoeffer, Viktor Frankl, and Eric Liddell. And still, many others are buried in the archives of history, including Carlos Montezuma, Bartolomé de las Casas, and Thomas Merton.

in a world dominated by other thinking styles. While Style V thinkers use more abstract thinking methodologies, they still maintain abstract ideals that are either firmly attached to a status quo (as in a Newtonian or heuristic perspective) or are in close proximity to it. They do not like straying too far beyond their accepted and established system, whatever that system might be.

Because of this, Style V thinkers can usually fit into society fairly well (with a few exceptions). Provable facts reign supreme with them; the how of a given thing is still of primary importance, but the why is almost as important. When Style V thinkers try to discover the hows and whys of a given problem, they usually use methodologies that control for variables in an effort to isolate specific truths. Once those truths have been tested, established, and isolated, then and only then do Style V people draw conclusions about reality. The key handicap to this method of thinking is that it runs parallel with Style III thinking, in that truths are isolated to specific situations instead of a universal whole. Style V thinkers represent about 10 to 15 percent of society.

Style VI - This is probably the oddest thinking style of them all. Whereas, thinking styles I through V primarily use linear methods to process information, connecting *A* to *B* to *C*, Style VI thinkers use multidimensional or quantum thinking methods. That is, *A* is probably not what we think it is, and its connection to *C* probably only exists within the constructs of a linearly organized world.[67] In many cases, Style VI thinkers are what a linear-thinking public would define as either odd or extremely eccentric, because much of what a Style VI thinker conveys to others is extremely complex. The inner workings and mental language used by Style VI thinkers usually renders its conclusions incomprehensible to a linear mind. Why are the Style VI thinkers so hard to understand? They produce multilayered and multidimensional thought experiments that do not follow linear paradigms.

---

67 We'll discuss linear reasoning in further chapters.

Furthermore, the processing of this information usually corresponds to deductive inferences leading to unrecognized associations between formally incalculable or dissimilar variables that ultimately lead to the discovery of universal truths. These thinkers realize the totality of humanity's perceptual constraints, so they seek further knowledge using methods that go beyond established systems and norms in order to open up new avenues for knowledge.

Style VI thinkers are the most advanced knowledge consumers and producers on the planet. On the other hand, they are often considered the most backward individuals in society, an odd paradox to say the least. Style VI thinkers are fiercely independent and chafe at self-limiting, self-imposed conventions and assumptions. For this reason, Style VI people do not blend well with society as a whole and are constantly misunderstood. Style VI people represent less than 1 percent of the general population.

- LINEAR REASONING -

In this section we will take a look at how a linear mind works and how it reasons through information.

Cognitive scientists know that human reasoning is based upon linear processes, linear in that we view and process information using linear projections or linear metaphors in order to conceptualize our world. For example, due to our finite natures, the limitations of memory, and living in a four-dimensional world, we come uniquely wired to think linearly, seemingly to ease our cognitive loads. Because of this, in many areas of life, we have assumed things as objective realities when in fact they are subjective deductions at best. Case in point: we have created constructs in our daily lives that place limits on things that are either too hard or too complex to fully comprehend in a linear fashion. Every human being on the face of the earth likes things that are simple to understand, operate, and control (all linear processes).

Linear reasoning is composed of perceptions that are categorized in progressions and patterns. And because of this, humanity always narrows its questions of knowledge, reality, and truth to fit within these limited perceptual constraints. Generally speaking, our questions are usually narrowed down to some form of measurement (sets, compartments, dimensions, type, time, weight, speed, and so on). This is the way that we grasp and understand the universe and attempt to qualify and quantify everything. But when it comes to things that can't be measured by our linear minds, we grope about as in a fog.

Linear reasoning is anthropomorphic by its very nature (producing, among other things, the anthropic principle) and is at a great disadvantage when trying to understand anything beyond the human sphere of influence.[68] Objects, metaphors, and places are held within certain bounds and are usually connected to time (e.g., the arrow of time) in some way. To quantify a thing is to know a thing: this is the complete essence of linear

---

68 The anthropic principle, by its very definition, is a form of self-fulfilling prophecy. Observed physical constraints are remarkably compatible with what's observed in the universe.

reasoning.[69] This way of thinking works well within our closed system, but it certainly is not the only way to reason.[70] For the sake of discussion, let's examine two fields of study that are considered complex and consisting of objective truths, but are in fact simplified linear constructs.

Mathematics - We believe mathematics to be an objective discipline because it has utility and function in the world in which we live. But even something that seems as clearly objective as mathematics has at its core an incompleteness that in turn highlights our ignorance of reality. Mathematicians such as Gregory Chaitin, Kurt Gödel, and Alan Turing have all pointed out the ghostly foundations upon which mathematics is built. The question needs to be asked: what assumptions do we carry as objective truths that are subjective variables at best? As it turns out, there are many!

Have you ever asked yourself why we use a math system based entirely on linearly ordered systems (i.e., the decimal, positional notations, sets, grids)? Why don't we use some other system of mathematics when calculating?

Most of mathematics is composed of mental constructs (formalizable linear progressions) with the goal of reducing or simplifying irreducibles (e.g., infinity, space, time) to the conceivable (axiomatic methods, empirical theories, Cartesian systems, logarithms, and so on). Almost every objective or empirical truth in mathematics is in fact a subjective creation. For example, numbers (including zero) and mathematical symbols do not exist in and of themselves; they have absolutely no objective existence outside of mathematics. This is an important point to remember in our forthcoming discussions of: empiricism, rationalism, determinism, free will,

---

69 Examples: Deductive reasoning relies on premises, rules, propositions, and standards. Inductive reasoning relies on observed particulars and patterns. Linguistic reasoning relies upon rules and conventions. All of these reasoning styles fall within linear schemas.

70 When does a system stop being a linear combination of micro states (a closed system) and become a part of a universal unified state?

and the senses. When we examine mathematical numbers and symbols, we are immediately confronted with a dualism. In the realms of a closed system, the numbers and symbols we create have meaning. But as soon as we step outside of a closed system environment, our created constructs may or may not have any relationship to the objective reality that stands beyond us (the open system). Mathematics operates in a closed system while appearing to be a discipline that has no limitations. This is sometimes hard to wrap our brains around, so let's see if we can conceptualize it by using several examples.

Linear reasoning would confirm a number and symbol theory that states 1+1=2. And from our experience as human beings, that formula is not only rational, it also works extremely well within our closed system. But from any vantage point beyond our own, we do not know whether anything actually equals anything else, or whether numbers themselves should be static or dynamic entities. There is never a way of knowing, via our senses, if two items that appear to be of equal weight, size, or shape could ever equal each other exactly. Even within the four dimensions of our closed system, we cannot say with any certainty that 1=1. No one has ever been able to physically make one item exactly equal another. Why? Because there's always the possibility of something beyond our comprehension being unequal or impossible to measure (i.e., infinity).[71]

---

71 Humanity has a great deal of difficulty with infinity. Every place we look and everything we touch has some connection to the infinite. In mathematics, strings of numbers go on infinitely. When we reduce numbers, we can go on infinitely. Beyond our universe, space can go on infinitely. When we view the subatomic world, diffusion of energy, transmission of information, and so on, the infinite or eternal is as much an objective reality as our existence. It is one of those realities that goes unnoticed by the senses but is explicitly known as a reality by way of rationalism. Can science or mathematics measure the infinite? Can we use our senses to comprehend the infinite? Should we exclude its reality because we can't measure or sense it? If anything, the infinite should be placed in our lexicon of known truths as a foundational axiom for objective reality. It is one of those keys to understanding objective reality that has been hidden in plain sight.

Thus, on one level, mathematics deals with the senses. One orange plus one orange equals two oranges. Once we move beyond our senses, we know (by way of rationalistic and deductive systems), that we do not have the ability to quantify each orange's subatomic structure to determine whether or not they are actually equal. In other words, the mathematical constructs may have absolutely no ties to the reality that supports our closed system. And yet—and this is nothing less than momentous—mathematics also touches on the objective reality that stands beyond the subjective constructs of our closed system.

To the casual reader it might seem like I'm trying to tear down or degrade mathematics. That is far from the truth. Mathematics is a field of study that sets humanity apart from all other species. It is nothing less than pure creativity, imagination at its best and aesthetic beauty par excellence! And because of this, humanity can actually reach beyond itself in discovering new avenues to knowledge that extend beyond our closed system: non-Euclidean geometry, superreal and hyperreal numbers, and infinite sets, to name a few.

That being said, in its purest form, mathematics is more art and philosophy than science. Mathematics can create a new form of art (imaginary numbers, complex numbers, Cartesian planes) that might have very little to do with objective reality. The real challenge is in separating the subjective inventiveness of mathematics from its objectified discoveries. When we look at mathematics from an open system perspective, it becomes immediately apparent that closed system mathematical constructs deal exclusively with approximations, which happen to work within a wide latitude of subjective assumptions. Mathematicians who have explored the underlying assumptions of closed system mathematics are keenly aware of its tentative and subjective underpinnings. They realize the subjectivity of its foundations because they have knowledge of, and take a wider view of, the open system truths that actually sustain our closed system.

Make no mistake: mathematicians and all who use math on a daily basis have an underlying faith that the system is valid regardless of its closed system assumptions. For example, the mathematical work that Newton accomplished is nothing less than brilliant and has served science and humanity well. From a closed system perspective, Newton was a genius, but when viewed from an open system, he just barely scratched the surface. Numbers alone will never reveal the reality concerning universal truths.

Science - I cannot think of any other field of study that has given humanity more concrete examples of closed and open system realities than modern-day science. Like mathematics, science deals with a dualism that is not readily apparent to the masses. But unlike mathematics (with the exceptions of theoretical physics and cosmology), science has severely restricted its access to knowledge because it always narrows its questions down (via linear reasoning) to three elements: measurement factors, degrees of observation, and the requirement for repeatable results.

One of the foundational building blocks of science is the postulate; that is, all laws of physics hold true for all frames of reference. The speed of light is a postulate. So, the question should be asked: do our postulates connect to any absolute truths or objective realities? Maybe, and maybe not. But are postulates taken as a fundamental and objective reality in the field of science? Yes! And herein resides the truth of the simplistic nature of science. In science, the narrowing processes, so crucial to finding specific answers, actually excludes vast amounts of complex information that have direct bearing on the given results. To base knowledge entirely on a self-limiting system would be to strip away its normative properties, and thus creating a pseudo knowledge based entirely on diminutive dynamics. Is this really what we want to center our lives around? For many millions of people, science is the foundation of all that is in the universe and should be considered as objectified reality in all circumstances. In modernity, a common refrain is often voiced, "I don't believe in anything but scientific proof."

This is a shortsighted statement to be sure, but entirely understandable given an education system that is dominated by Enlightened belief systems. In a very real sense, we have placed an inordinate amount of trust in science.[72]

Let's step back for a moment and ask ourselves four critical questions about science:

1. Does the fact that the majority of observable events (past, present, and future) happen outside of human purview have any bearing on discovering truth?

2. Can science repeat *everything* that has a bearing on objectified reality? Another way to look at this, do non-repeatable occurrences have validity since they can not be repeated by scientific experiment?

3. Can we afford to assume that "quantifiable" data and methods are completely objective when we have such a limited knowledge of the universals involved?

4. And (this would apply to mathematics as well), how are ideas of epistemology, timelessness, infinity, causation, and absolutes handled?

These are all perfectly reasonable and fair questions to ask about science. Because science narrows (simplifies) its questions down to observable measurements or scientific axioms, it therefore can control for particular variables (obviously not all variables), making deductive and inductive

---

72 This is directly related to Enlightenment philosophy, education, and tradition. As we have shown already, proofs in one realm do not equal reality in all realms especially when the proofs remained bound by a closed system.

reasoning possible.[73] Such simplistic narrowing techniques are evidence that science is still in its early stages of development.

When it comes to things that cannot be simplified due to their extreme complexity or the excessive number of variables involved (and their interrelatedness with each other), the scientific method falls short. And, of course, this poses intractable problems when dealing with such concepts as existence (why anything exists), space (connections with matter, time, and dimension), infinity (what supports or goes beyond the universe), time (illusory, directional, or relative), and, finally, the nature of beginnings (time Alpha) and endings (time Omega). Linear methods of thinking and processing information will always have pitfalls, whether in mathematics, science, philosophy, or any other field of human endeavor.

The systems we use today places immense worth and confidence in science. This is nowhere more evident than where technology is concerned. Science that is geared towards technology happens to be one of the main drivers behind the capitalistic system. So science not only fascinates us, but also occupies a central place in our economy. In many ways, science is placed on a pedestal from which it can do no wrong. Needless to say, placing one's faith in a single system (or even a couple of systems) creates an environment of bias in which truth is usually the sacrificial lamb.

As we did with the topic of mathematics, I'd like to conclude our discussion on science with three things to remember. First, science is still an extremely useful tool; to deny its usefulness based on its structural restrictions would be excessive and unreasonable at best. Secondly, science, like mathematics, reflects a faith in unknown fundamentals. And thirdly—and this point will have a great deal of relevance in later discussions—science creates its own pseudo absolutes (e.g., postulates, theories, axioms) while denying absolutes in other realms of reality. This bias is prejudicial and

---

73 An important point to remember; in science, theorems only prove what the axiom states. The axiom can be incorrect and still prove valid! In other words, the axiom is only true to itself. Provability does not necessarily equate to fact or truth in science; it might, but then again it might not.

irrational to say the least. Suffice it to say, like mathematics, science alone will never reveal the total reality concerning universal truths, for that we will need much more. Now let's finish our examination of linear reasoning by contrasting it with another model, that of quantum reasoning.

– QUANTUM REASONING –

"And the bay was white with silent light
Till rising from the same,
Full many shapes, that shadows were,
In crimson colours came."

Samuel Taylor Coleridge

For most of us, the idea of thinking quantumly is beyond comprehension. Why? Because there is nothing for the linear mind to grasp in quantum reasoning. To put it another way, in quantum reasoning, the targets (the questions and the answers) are connected but almost always stand outside of spatial boundaries and time components on several different fronts (differing planes, multiple dimensions, convergent systems, and so on) and are usually removed from things that are never quite what they seem. To the linear mind, quantum reasoning can seem like pure nonsense. And that's probably what you are thinking right now. So let's first look at three examples to help us understand quantum reasoning and reality.

First, in Isaac Newton's era, space and time were considered static, absolute, and unchanging. Then along came Mr. Einstein, whose' ideas turned the world upside down. Even to this day, most people still do not comprehend the implications of Einstein's discoveries because his theories and reasoning methods dipped into another realm altogether, that of quantum reasoning. Einstein took Newton's ideas of space and time (simplified, reduced, and quantifiable) and, by thinking about them quantumly, showed them to be moving, dynamic, and relativistic targets. Then he merged them into a more complete macro-entity called the space-time continuum. Separate one from the other, and both cease to exist, just as if you tried to separate your brain from your body, both would lose significance. By reasoning quantumly, Einstein showed that things are not as linear as we would like to believe. In the case of space and time, each is relative to the other and interacts with the other in dynamic ways.

100

Secondly, linear reasoning tends to reduce things to their simplest forms. If our mind can reduce information down to two or three variables, objects, or themes, then we are happy and contented people. But quantum mechanics has proved that things can be very different from what we comprehend via our senses, observations, and measurements. In the world of quantum mechanics, chaos appears to prevail *until we try to comprehend it*. Odd, to say the least! In quantum physics, when we measure an electron's position, we are not measuring anything that could be remotely considered objective or preexisting. The very act of trying to measure the electron's position in time and space (imposing temporal reality upon a transcendent open system) has resulted in two significant insights into objective truth and reality: first, that the open system is very different from what the linear mind can grasp, and second, that we will never be able to reduce all the information necessary in order to completely understand the objective reality behind our subjective creations.

And finally, how do we define such words as *beginning* and *ending*? Webster's dictionary offers a traditional, linear definition of each that is satisfying to the majority of people. Viewing these words from a very basic quantum approach, however, you would realize that the two words are essentially one! To have a beginning is also to have an ending, and to have an ending is to have a beginning. You cannot have the one without the other. One word defines the other, and they both constitute a unique whole (combined with many other things). Separating one from the other makes both cease to exist. To further complicate this, the quantum approach asks *why*? Why does there have to be a beginning and ending? Is the idea of beginning and ending another human construct, a Newtonian view that believes such things real when in actuality they live in an infinite state of being? This is quantum thinking on a very simplified level. Quantum reasoning addresses ideas or concepts that cannot be completely described using linear methodologies.

Thus, from a quantum reasoning perspective, we know that linear thinking methods do not have the monopoly on knowledge, reason, or reality. As linear thinkers, we must accept that our way of thinking is subjective at best and totally irrational at worst. Also, our inability to understand a given concept, definition, physical property, transcendent happening, or other factor does not mean it has no value or a role in the grand scheme of things. From a quantum perspective, to value or devalue one thing is to ignore the interdependence and interconnectedness of all things. There is a matrix to life that connects everything in some form, and that in turn conveys an immensity and complexity beyond our finite and linear reasoning abilities. Linear reasoning places us in the position of building an artificial reality that may have very little if anything to do with universal truth and reality. We think we know quite a lot, but in actuality, we know next to nothing. It is only in that place where we truly comprehend our ignorance, no matter how smart or educated we think we are, that authenticity, honesty, and the ultimate truth can be found and comprehended. It's a lowly beginning, to say the least, but it's a great beginning nonetheless!

Fences

It was a winter's day in 1980 when I met the most remarkable man. I was living in New York City at the time, and I had gotten into the habit of taking a daily stroll through lower Manhattan. I entered a neighborhood park and noticed a man sitting by himself at the far end of a bench. I walked over and sat down at the opposite end, leaving a good space between us. We did not speak to each other at first; all was silent except for the occasional gust of wind that sent dried leaves scurrying about. The sun was shining, but no warmth came from it. There's never any warmth in New York City during the winter, and yet I always found it enjoyable to bundle up for a brisk walk, followed by a cup of hot coffee at my favorite café.

After taking my seat, I studied the man at the opposite end of the bench, inconspicuously of course. He had a brown cap upon his head and grey wisps of hair poked out from it here and there. A thick woolen coat was wrapped about his body, and a black scarf enveloped his neck and chin. The man's face was worn and wrinkled, with many age spots. It's odd the things you remember, but I remember looking at his left ear and realizing it was huge. I suppose his right ear was just as big, but just seeing one side of his face, I had to imagine it was so. By the man's dress I guessed he was from Eastern Europe. These old-country Europeans always seemed to dress in similar ways: black shoes, drab dress pants, and almost always a solid-color collared shirt that buttoned up the front.

When the man noticed me studying him—so much for being inconspicuous—he turned towards me and smiled. I smiled back. We greeted each other and talked about the day's weather. The man introduced himself as Peter and, as I had guessed, he was from Eastern Europe, Hungary to be exact, and yes, his right ear was just as big as his left.

I discovered a long time ago that whenever I meet someone, especially from an older generation, our conversations end up being remarkably fascinating. I've learned a lot about different people's experiences and histories by just sitting back and listening with a receptive ear, an open heart, and an employed mind. Peter's story was not just remarkable, but uniquely uncommon, poignant, and illuminating on many fronts.

After we talked about the superfluous for a while, I started asking questions about Peter's life in Hungary. Peter told me about his boyhood growing up on a farm, about his mother and father and brothers and sisters. He told me about the little town they lived by and all he could remember about growing up as a boy and young adult. Soon we were talking as if we were old friends. At some of his memories we would laugh, and at other times a palpable sadness filled the space between us. I soon found out that Peter was a survivor of Auschwitz, as in the concentration camp established in Poland by the Nazis during World War II. Auschwitz was just one of many camps in which millions of innocents were starved, tortured, and put to death, all in the name of Nazi dominance and superiority. As one might guess, the conversation took on a whole new turn. In a quiet and unassuming voice, Peter began to tell me about Auschwitz. I will not relate all of the horrors he told me about, most of which have been documented and verified in other places, but one of his experiences bears retelling, if for no other reason than revealing those hidden mirrors of the soul.

Peter said there were days he could hear music and laughter floating through the air of the camp. The first time he heard this music, he thought his mind was playing tricks on him. The music brought back fond memories of friends and family gatherings, times when life was good.

As the music floated upon the breeze, Peter looked around him. Those memories produced by the music quickly faded in the stark reality of his situation. He looked around: camp prisoners moved as silent wraiths, more dead than alive, with wasted bodies and hollow eyes, all waiting for the end to come. Even the ones who had died still haunted the camp as grey ash and black soot falling from the sky.

Eternally burning, the crematoriums consumed the vestiges of the body and the essence of what once was good. Men, women, and children were murdered by the Nazis and then burned like so much garbage. What was left of each soul was carried up through flumes and smokestacks, eventually becoming ghosts upon the wind.

No, Peter knew his mind must have been playing tricks on him. He was in a prison camp; there was nothing in Auschwitz to hear but the death knells that called all to the fires. Music was reserved for happy places and happy times; Peter knew there could be no music in a place such as this.

As Peter started to walk back to his barracks, he heard it again, unmistakable this time, faint, but there nonetheless. It was indeed women and men laughing as polka music drifted upon the air. Peter even thought he could smell the aroma of sausages cooking—oh, how that made his stomach churn and mouth water.

Peter could not believe it. Where in this camp could laughter be coming from? He looked around to discover the music's origin. It seemed to be coming from another compound within the camp. Barbed wire and electric fences separated him from the music, food, and laughter. Peter's emotions overwhelmed him. Tears streamed down his face as he realized his fate. So close, so tempting was redemption, and yet it could not be reached.

As the days passed and he compared notes with the other prisoners, he found out that the music and laughter was coming from the German guards' quarters. Peter found out that the camp guards and officers, along with their wives and girlfriends, often held parties to alleviate the stress of operating the camp. Many times Peter tried to reconcile the madness of the situation. How could people entertain themselves, eat abundantly, and be happy while being surrounded by wretchedness, deprivation, and death? Peter told me that this realization nearly broke his mind and spirit, as a bullet or starvation never could. This pretty much ended Peter's tale.

The next minute Peter told me he was getting cold and that it was time he headed home. We stood, shook hands, and said good-bye to each other. I thanked Peter for sharing part of his life with me, and he said it was a surprise that someone actually wanted to listen. Peter then turned and walked out of the park, never looking back. I watched him go as the shadows of the city closed about him, and that was the first and last time I saw Peter, but not the last time I thought of him.

Peter's experience in this regards was one of the most poignant reality checks I've ever had. To think just a few fencerows away from the merrymaking, a whole population was suffering and dying. I could imagine Peter and the thousands of other souls in Auschwitz standing in the prison compound, smelling and hearing the celebrations, and I'm sure occasionally seeing groups of people who lacked for nothing while they lacked for everything.

What is it in us that allows such a hardness of heart and callousness of spirit that we're willing to ignore others in their hour of need? Is there really a valid excuse for such actions or inactions?[74] One wonders how society picks and chooses whom they will help and not help, whom they will and will not allow to live. Have we changed, evolved, or advanced in how we treat others today? As the rich nations of the world enjoy their plenty, the majority of the world looks on through fences we've created, maintain, and protect. The real question is and has always been, do we do what is right in our own eyes, or do we do what is right according to universal and absolute truths?

In the next part of this discussion, we are going to shift gears a bit and delve into why we do the things we do, and why the majority of society, mostly Style II and III thinkers, think and act in certain ways. And to a greater degree, we'll ask why the world is so chaotic and dysfunctional today. We have already touched on this subject in previous discussions, but now we will dig under the surface in an attempt to discover the deeper issues that block our understanding of universal truths and quantum reality.

Chaos and Dysfunction

If this were a democratic process and you were to count the votes on whose reality is the correct one, the overwhelming majority of votes would be cast by Style II and III thinkers. Style II and III thinkers are by far the majority of thinkers on the planet. Their reality dominates every sector of life. This group includes bankers, executives, laborers, academics, rich

---

74 If we use a contemporary definition of truth (i.e., pragmaticism), then the answer to this question is an unequivocal yes, so we do have a pragmatic excuse and reason for allowing injustices to happen. In the case of the camp guards, had they not done as they were ordered (to beat, starve, debase, and exterminate), then they could be put to death themselves. Is this pragmaticism at its best or worst? It is neither, because pragmaticism is simply a justification used to nullify a universal truth, in this case, respect for life, human life.

people, poor people, and just about everyone in between. In fact, it would be much easier to discuss the people who do not use Styles II and III thinking methods than to fully address how Style II and III thinkers dominate society. But the fact remains, because this group influences everything from the most mundane to the most critical aspects of life, they demand our attention. In order to introduce the subject of Style II and III thinkers, let's begin with the concepts surrounding loyalty.

Loyalty seems to be one of those coping mechanisms that is hardwired into the human psyche. The most obvious reason for showing loyalty to others is in the building of interpersonal relationships and group cohesion. Showing loyalty helps us navigate situations involving cognitive dissonance and heuristic methodologies. Style II and III thinkers show an insatiable drive to please and appease others in order to be socially accepted (by family, friends, social groups and organizations, and so on). This is a good value when it comes to commitment to a truth or a justified cause, but it can become problematic and dysfunctional when universal truths are trampled upon by heuristic thoughts (a type of linear thinking) and the valuing of unconditional loyalty above universal truths.

The problem with unconditional loyalty and commitment is the ease with which the good can turn to bad and then to dangerous in a very short time. When loyalty is won at the expense of something else (i.e., a person, a society, an ecosystem, etc.), then that loyalty becomes a monster of one kind or another. All of us can think of examples in which peer pressure, wanting to fit in, or blind loyalty to a cause or a person has caused humanity to act abominably.[75]

Because of Style II and III thinking, political, economic and social systems tend to be corrupted in one way or another. This is primarily due to misplaced loyalties that compromise ethical and moral values (i.e., those values based upon universal truths) in the pursuit of maintaining group

---

75 Nazi Germany, Stalinist Russia, Mao Zedong's China, Pol Pot's Cambodia . . . the list is endless.

cohesion.[76] Blind loyalty corrupts more things than most people can imagine and is a leading cause of many of the problems we experience today. Heuristic methodologies, groupthink, tribalism, and self-sustaining biases (all of which are particularly entrenched in Style II and III thinkers) are the results of taking mental shortcuts where none should be taken and giving unconditional loyalty to another person, group, or cause to gain acceptance. Wars, political misconduct, corporate misdeeds, ethnic conflicts, and religious wrongdoings are all the results of trying to sustain misplaced loyalties. These things are bad enough in and of themselves, but the darkest side of heuristic methodologies and misplaced loyalty is the eventual change in the way a person thinks and acts.

A person can enter a group with altruistic aims and be initially alarmed when the group does something that goes against a moral value or universal truth. But because of our drive for simplistic answers—that is, heuristic or reductionist ways of thinking—and in order to be accepted as team players, we create cognitive tools to help us cope with the internal conflicts between universal reality, complexity, and the group's accepted norms. These cognitive tools come in many forms, including suppression, denial, self-justification, and pragmatism. When loyalty is established, whether the person is aware of it or not, they've just become a slave to another's desires. This enslavement could be to systemic norms or to a single, highly esteemed person; it could be to the norms of a small group (one's family, friends, members of a club), or it could be to the norms of

---

76 We are not talking here of those moral or ethical systems that ignore or deny the self-regulating axioms that provide humanity with health, stability, and life. For example, the moral standard that stealing from another is wrong would be a universal truth that was, is, and will continue to be an axiomatic standard for humanity, a universal truth that holds over distance and time. Most moral and ethical truths have been challenged on the grounds that they do not apply to all people, in all times, and in all places. But if you'll recall our discussions concerning truths and dualities, in order for there to even exist a grey area to debate, there also has to be a universal set of absolute dualities that bracket and give some form or substance to those grey areas. The bottom line is that we cannot debate the grey areas without first having some absolute truths from which those grey areas emerge. It is the universal truths that concern us here.

a very large group (a corporation, a religious organization, the military, one's political party, or a worldwide organization).

Because Style II and III thinkers make up the majority of people, their heuristic thinking and any misplaced loyalties (those perceived norms) eventually become intimately and fundamentally connected with how they think and act, who they want themselves to be, and in the end, will believe themselves to be.

To Be or Not to Be, That Is the Quantum Question

For the most part, Style II and III thinkers must create and wear a personalized mask, a persona, to maintain acceptance. In most cases this created persona is an emotional device that helps to elevate and protect us from the hidden realities buried deep within our subconscious worlds, worlds of insecurity, pain, self-centeredness, fear, and shame, all of which are emotionally based platforms for the construction of a false reality.

Another way to look at this is from the perspective of protecting ourselves from others or from hard truths that we do not wish to accept or believe. We are fully conscious of the masks we wear in these situations, whereas the masks we wear to hide our inner selves are, more often than not, created and worn on a subconscious level. Either way, the persona is always emotionally based and always used to defend against or to deceive the world, and in many cases it's used to deceive ourselves.

A person wears this mask to convey what they wish the world to see or what they believe themselves to be, or not to be.[77] The sad fact is that most everyone you meet wears a mask of some kind, which in turn means the person is being less than authentic with themselves as well as you. Most of us create a persona that puts as much distance between us and our pain or inner liabilities as possible. This created persona gives us a measure of comfort and creates the illusion that we are in control of those

---

77 Examples would include being considered strong, sophisticated, smart, tough, savvy, funny, trendy, or in vogue.

less desirable or socially distasteful elements that have become a part of our lives. If we can control how others perceive us, and if we can control how we perceive ourselves, then we can sincerely believe that the undesirable side of our inner selves, or the pain that others cause us, has been vanquished or at least fairly well contained.

I've witnessed people with learning difficulties who, out of fear of being perceived as stupid, go to extraordinary lengths to create a persona that is both smart and in control. In many cases, these people may even end up as very successful entrepreneurs, military leaders, business executives, athletes, entertainers, gang leaders, and so forth. Other individuals may wear a sanctimonious mask of disdain for others in order to feel better about themselves. These people elevate themselves above the crowd, judging everyone to be a lesser being.

I've also seen people with disgraceful pasts who have worked very hard to create a persona that is well respected or loved in their chosen community or group(s). These people seek to create a persona that is beyond reproach. I've witnessed people who have been extremely self-centered who immerse themselves in religious organizations or volunteer pursuits that advertise themselves (to the outside world) as virtuous and self-effacing while all along they are feeding their egos.

As a side note, there is a constructed persona that comes about almost naturally, but is nonetheless counterfeit. One sees this particular persona in cultures or households where there's a lot of violence or abuse. It shows the world a blank or uncaring face; it is a persona that has been forced upon the person as a means to survive the extremes of psychological abuse, sexual exploitation, suffering, bloodshed, or hostility. This persona is a mask of protectionism in the extreme.

In most cases, when these finely tuned and constructed personas are challenged, threatened, or on the verge of being exposed for the created edifices they are, the person who is in danger of being revealed usually goes on the defensive (often becoming angry, argumentative, self-justifying, blaming, or rationalizing). This person might not even realize that his or her subconscious is coming to the forefront to do battle, or understand why their usually calm demeanor is suddenly emotional and explosive. The real reason behind the defensiveness is that a falsity, both conscious and unconscious, has been revealed, and one that the mind will defend against at almost all cost. Indeed, we like to believe the illusions we create about ourselves. Many in society tell themselves that they're smart, intelligent, good-looking, powerful, in control, and even intrinsically good, while ignoring the fact that they live in a delusional and remarkably inauthentic state of being.[78]

The use of heuristic methodologies, granting of unconditional loyalty, and showing unreserved obedience to a system does not usually happen overnight. It is a gradual process that develops over many years, literally from childhood on. This process molds our subconscious and conscious views on life. It socializes us to a system that doesn't necessarily align with absolute reality. This acceptance of norms takes more permanent form as we begin to make adult choices on how to live our lives and how we want to be perceived by the world. To a very large degree, Style II and III thinkers establish their self-worth, self-esteem, and self-identity upon the people, groups, or systems to which they attach themselves. What others think or expect of them becomes the driving force behind their thinking and actions in life.[79]

The attachments Style II and III thinkers have to fitting in and obtaining the approval of others may give them the illusion of freedom, but in actuality their freedom is severely circumscribed by the very people or groups to which they offer unconditional alliances. Unfortunately, Style II and III thinkers rarely, if ever, challenge the accepted norm, because doing

---

78 People who create their own personas can typically (though not always) see and detect personas in other people. It is remarkable that other people's personas can be so apparent while one's own remains hidden to a fault. I speak from direct experience: I never knew I had made a false persona until a situation presented itself to me several years ago. A hard reality—I ended up being someone other than my persona wanted me to be, a person I was not even sure I knew.

79 This is a type of social codependency.

so would mean being labeled as disloyal, rebellious, or untrustworthy, labels which are anathema to a Style II or III thinker. These and other undesirable monikers, of course, are meant to reestablish control over a person's thinking or behavior. Unfortunately this method of control works very well with Style II and III thinkers.

Demosthenes once said, "Nothing is easier than self-deceit." Indeed, it is much easier to go with the flow of the group, the organization, or the system, than to step outside of that norm and think independently and objectively. Because of this, loyalty seekers (mostly Style II and III thinkers) live in a self-rationalized reality that never pushes beyond their chosen set of norms or loyalties.

Most Style II and III thinkers do not realize that their loyalties, actions, or inactions are causing harm and distress to others. For the most part, Style II and III thinkers believe they are doing the right thing and acting as responsible people, while at the same time missing the opportunity to look beyond themselves to see a broader, more quantum picture of life. As the above encounter with Peter revealed, the fact that 80 percent of the world lives in poverty is totally missed or marginalized by Style II & III thinkers living in the wealthy nations of the world. The poverty surrounds us. Men, women, and children are dying every day because of it, but the majority of Style II and III thinkers in today's world simply ignore it, much like the Nazi guards once ignored the hundreds of thousands dying around them.

Once the mind makes the subconscious switch to a group (or system for that matter, e.g. the Republican or the Democratic political system in the U.S.), from that point forward the person has given his or her power over to the group, and woe to the person who goes against or steps away from the group's standards.[80] Once the mind makes the switch from objective reality to subjective reality, the group can commit monstrous acts

---

80 If you want to see the power and control given to the group, just step outside of its norms and watch the power behind the machine. When you actually start thinking for yourself and basing your actions upon absolute reality, you're sure to draw abuse by the system's agents of ignorance.

or seemingly small acts against those universal truths that give us life and meaning, and still they will be justified by the group's acceptance of whatever subjective reality they've created. To further clarify this fact, a few examples may be in order:

- A person, prior to joining the military, has the value of protecting innocent life (i.e., noncombatants). After joining the military, he or she realizes that the larger command structure routinely bombs civilian targets, killing innocent people. There are many justifications (pragmatic answers) given for these acts, which will relieve an individual's cognitive dissonance while at the same time helping that person to be accepted as a team player. But by accepting these justifications (i.e., consenting to a subjective reality), the person totally avoids the objective truth of the situation in which innocent civilians are still being killed.

- Prior to being hired by a business or corporation, a person holds to the values of equality for all. After joining the corporation, the person is assigned to a team tasked with increasing corporate profits. The team focuses its attention on a manufacturing plant located in Kenya. The team knows that most of the employees at this plant live at or below the poverty level; with that knowledge, the team compels the workers to work longer hours for less pay. Since these employees live in poverty to begin with and have little choice when it comes to feeding their families, they are forced to accept the company's demands in order to maintain some form of income. Many justifications might be given for this coercive act, but none of them change the absolute truth of the situation: the Kenyan employees are not considered equal to the company's European and American employees. The objective truth is that one set of employees struggles to survive, while the other set owns cars, patronizes restaurants, and vacations at Disneyworld.

- A newly elected politician begins a career with the values of democratic justice for all. To help with the ensuing political campaign, he or she accepts a large cash contribution from a consortium of oil companies. This politician is eventually elected, and as time passes he or she is wined, dined, and treated as royalty by the oil companies that funded the successful election. Two years into public office, the politician discovers this particular oil consortium has collaborated and colluded in a price-fixing scheme to fleece the common consumer. Since this politician is friends with and accepts financial and personal assistance from this oil consortium, he or she decides to look the other way and does nothing to change the price-fixing scheme. Once again, there are legions of justifications for ignoring the situation, but the objective reality of the situation is that consumers are being swindled.

Unconditional loyalties to friends, organizations, causes, countries, or even families are sure mechanisms for creating dysfunction.[81] Is there a solution to the problem of misaligned loyalties, groupthink, and the acceptance of subjective realities? Indeed there is. We all have the responsibility for maintaining a sense of conditional neutrality when attaching ourselves to persons, groups, or causes.[82]

On a subconscious level, Style II and III people try to walk the fence between their ideals of doing what they believe to be the right thing and of spending vast amounts of time and money on fitting into their chosen set of norms. This is particularly true in the wealthy, industrialized societies that have the time for indulgences. Because of this, Style II and III thinkers like the *idea* of being charitable, honest, humble, and sincere, not necessarily because they are these things or live that kind of lifestyle, but

---

81 This whole dynamic surrounding loyalty and social acceptance borders on pathological codependence, which is typical of Style II and III thinkers.

82 This piece of noteworthy wisdom can be used in many situations and on many fronts, and it goes a long way towards healing the wounds of our world.

because if they are thought to possess these ideals, others will love and respect them based upon the illusion alone. They do not need to spend time and money helping others; as long as their peers *think* they are a good person, then they must *be* a good person. [83]

Please know that when I mention these things, I judge myself as well as Style II and III thinkers. Everything I have mentioned so far, and will discuss further, applies to me as much as it applies to other people. I do not stand in judgment of others actions or inactions without also standing in judgment of myself. Even now, I know that I'm living in excess because of the things I have allowed to clutter my vision of what's important in life. I can cut back tremendously on how I live simply by scaling back and getting rid of all the superfluous things I have or do.

So what is the main point to remember? The main point to remember is two fold. The first part is that of humble self-examination. The easy path is that of indignation at being judged deficient. My dear friend, we are all deficient in one way or another, but the quintessential question is, what do we do with those deficiencies once we know about them? Do we continue to live as we always have lived, or do we change for the sake of life and the planet? The second part to this equation is to consider what our deficiencies may symbolize. Give that last statement time to sink in and develop, what does it mean? Our deficiencies symbolize many things on many different levels and once we catch the significance of their symbolism, we can mature and move on to greater understanding and wisdom. One of the core symbols to be noticed has to do with the whole notion of compartmentalization. If we are people who desire to live complete/holistic lives, that is, lives that are lived to walk in harmony within the Matrix of life, then living that life will eventually mean the death to compartmentalization. In other words, I may be a well paid white collar professional who

---

83 Some Style II and III thinkers do live within their means (not in excess) and are very charitable, but they are not the majority.

spends a lot of time and finances on helping other people, but at the same time drive a luxury car, live in a luxury home, and spend a great deal of time and money on entertaining myself and others. This is a compartmentalized life in that my commitment to living a life within the bounds of true reality is flawed, i.e. I still spend resources on the superfluous while children are still dying of malnutrition. With that said, let's drill down even further, keeping in mind not our indignations, but the symbols that Style II and III thinkers represent and our ultimate response to those symbols in light of the quantum reality that surrounds us.

There's no doubt that Style II and III thinkers have a psychological compulsion and sociological addiction to the status quo (almost always a created system based upon half-truths and non-reality). This compulsion is a disproportionate obsession with maintaining one's image in direct opposition to truth and authenticity. In other words, objective reality is replaced by subjectivism.[84] Welcome to the real, not so real world of subjective reality, we are literally surrounded by it, a created system that is anything but a system based upon universal truth or objective reality.

Let's take a moment to talk about today's status quo. The most widely known status quo system today is popular culture. Popular culture tells Style II and III thinkers that they are smart, wise, and even sophisticated because they do what everyone else is doing, they think like everyone else is thinking, and they always act accordingly. When this type of group-think occurs, reality and truth are easy to define, because everyone else is acting and thinking in similar ways.[85] This type of psychological compulsion manifests itself quite readily in the desire to keep current with the latest and newest trends. For example, if a VIP comes up with a new design, a new way to entertain, or a new technology, then it's almost immediately embraced by Style II and III thinkers in their attempt to be viewed as

---

84 This falls within the parameters of social network theory, which attempts to describe how individuals reduce their cognitive dissonance (mental discord or disharmony) that is associated with group and network dynamics, in other words, the things people will do in order to appease the group and reduce their own internal conflicts.

85 This also becomes a self-fulfilling prophecy for self and group justification.

up-to-date, smart, cutting-edge individuals.[86] Marketers (of news, entertainment, politics, sports, media, or any other product) create and promote popular culture to be used as a tool for marketing their wares. If the marketers can manipulate people to stay within the circle of pop culture, then they can sell anything by providing everything that gives people the illusion of being in life's game. Obviously, everyone wants to be considered innovative, intelligent, and in vogue—it is a type of loyalty to the majority. This commodified mentality keeps billions in a state of self-centeredness and immature longing to stay current with whatever is being peddled as cool, clever, smart, and fashionable.[87]

On the whole, I feel safe in saying that 99 percent of people want to be liked by their peers. This desire to be liked by others is an axiomatic force of the human condition. True, thinkers of all styles wish to be liked and accepted by their friends and associates, but the difference is in how much importance a person assigns to the task of being liked, esteemed, and assimilated within a group or system. Style II and III thinkers place an inordinate amount of importance on how they are viewed by others. Instead of basing their decisions on objectified reality, they use their feelings and emotions as de facto measures for defining reality. Hence, reality becomes emotive and personal instead of cogent and impartial.

In the final analysis, the perception level of Style II and III thinkers becomes distorted by the outside influences they allow to dominate their lives. Because of this, most Style II and III thinkers fail to develop beyond themselves. In essence, they have been sold a bill of goods that enslaves

---

86　The VIP is usually a person highly esteemed by Style II and III thinkers. This person may be a renowned scholar, entertainer, financial investor, entrepreneur, activist, or other authority, just as long as they subscribe to and endorse the establishment—that is, Style II and III ways of thinking.

87　Today in the United States, if you speak a certain way, dress a certain way, or are cynically or sarcastically funny, then you are obviously in the know, in the tribe, and thus socially acceptable.

them in a world of irrelevant thoughts and behaviors.[88] It is heartbreaking to stand on the outside and look into the world of Style II and III thinkers. They settle for so little, and yet they have the potential to go far beyond the mediocrity of fitting in.

It is truly ironic that some of the world's smartest and most educated people still process information as Style II and III thinkers. These individuals would be easily recognizable as some of the greatest men and women in technology, science, economics, business, and entertainment. One of the greatest misconceptions of our age is the belief that intelligence somehow equates to wisdom. Some of the smartest people on the planet are also some of the most foolish. The possession of intelligence does not and never has translated into the possession of profound understanding. This differences between having intelligence and wisdom manifest itself in many ways, let's look at a few of them:

- When we worship those who are considered elite, smart, entertaining, or wealthy (e.g., the MIT grad, the inventor of the latest gadget, the comedian who makes us laugh, the sports star that wows us with his or her athletic abilities, the financier who makes billions), we have to ask ourselves: could we ever take seriously the advice of someone living in poverty in the slums of Calcutta or on the streets in New York City? Most Style II and III thinkers would not take advice from the poor and destitute of the world, because doing so would violate the unspoken law that says smart people are not destitute specifically because they are smart, shrewd, and wise. Do you sense any discrimination here? A Style II or III thinker might say, "We have compassion for these people, but taking advice from them, learning from them—let alone changing our lifestyles and

---

88 The meaningless and inconsequential things of life that somehow become subjects of intense focus include: what sports team is the best, what musician or artist is the greatest, what kind of sunglasses do I wear, what wine should be served with pasta, what kind of tattoo should I get, what kind of car or motorcycle should I be seen driving, and so on.

ways of living and thinking because of them—is unrealistic and naïve." The belief that you are well off due to intelligence or a strong work ethic is fallacious. Followed to its logical end, this attitude holds that a person living in poverty must be one of, or an amalgam of three things: foolish, unintelligent, or lazy.

• How many truly impoverished people are asked to speak at meetings of the World Trade Organization (WTO) or Congress, to lecture at Cambridge, or to appear at the glitzy conferences of the sophisticated and elite held around the world? Were this impoverished person allowed to speak, how many of us supposedly elite and intelligent people would listen, let alone respect a person who speaks in an uneducated way, who has achieved nothing in life beyond survival, or who might be malnourished and disfigured? Since there's no smart value, inspiration value, or entertainment value in bringing the impoverished to the stage, they are ignored even as we pay them tribute with words, pennies, and the occasional gift of our precious time.[89]

• Could we ever have a government administered by the poor? Would public servants consent to live as paupers so that wealth accumulation loses its stranglehold on the wheels of government? The answer is no, because Style II and III thinkers deem wealth to be the fuel that powers society. "We cannot have poor people leading a nation because they are uneducated and stupid." While it's true that you'd never hear a smart person utter such a statement, it's also true that the actions of smart

---

89 This also applies to how we view and treat people who learn differently than we do, who work in low-paying or undesirable jobs, and who hold differing views than our own. It is always sad to see a person being ignored or snubbed by another who believes himself or herself to be smarter than average. The arrogant never realize how much they proclaim their own foolishness when they ignore the seemingly dimwitted or uneducated of the world. Obviously, if we know how to be sarcastic and demeaningly funny in our judgments of other people, it's self-evident that we are smarter and more perceptive than the fools we dissect piece by bloody piece.

people continually confirm the statement's validity. Few consider that a person without money might have the understanding and wisdom to do what's needed in a world obsessed with profit, power, and entitlement. For centuries we've had all the so-called smart, intelligent, and well-heeled people leading our governments. For the most part they have failed us and the planet we call home. Ever wonder why?

For a person to believe they have more value or are entitled to more, simply because they believe they're smarter or better than the millions never granted the same opportunities is preposterous to say the least. If someone in India, China, or Africa has the same intelligence that I do or even the potential for it, but I happen to live in a wealthy nation and therefore have the means to acquire an education and a large income, does it justify the attitude that I deserve more benefits while the rest of the world deals with famines, wars, and poverty? If you take a moment to actually place yourself in the role of someone struggling to survive, and if you are completely honest with yourself, you cannot help but see the discrimination that is taking place today. It's a discrimination that turns a blind eye to suffering and tries to justify its pleasures and entitlements at the expense of others. The objective reality of this situation is often overlooked by those who have plenty, but is painfully clear to those who have nothing. This insidious form of discrimination seems to be beyond the grasp of those ostensibly smart and intelligent people, and yet it is plainly evident to anyone with even a modicum of understanding, including those without much money.[90]

---

90 To clarify, the educated, the well-off, and the professed intellects of the world know and typically acknowledge the dispossessed of our age. These people are intelligent enough to know that poverty consumes the globe, but have not the wisdom to change their lifestyles in response to that knowledge.

The absolute truth of the matter is that we all share the same planet, and we are all human beings.[91] Today, the United States consumes approximately 40 percent of the world's resources while representing a mere 5 percent of the world's population. As our brand of capitalism spreads around the world, and as China begins to dominate the world stage, seeking ever more resources for cars, homes, electronics, electricity, and so forth, are we still willing to stand by as the poor of the world are used when convenience dictates, then jettisoned when they become problematic to the business model? Is there any justice or moral high ground in keeping certain institutions and businesses in power, to reap ever-growing wealth and dominance for themselves at the expense of humanity and the planet?

Capitalism, Quantum Reality, and Ground Zero

Our discussion will use the ideas behind capitalism to interface with ideas that complete a much larger picture of reality. This approach is rarely used, because it is foreign, uncomfortable to contemplate, and extremely intimidating when action is required. Indeed, it is much simpler to discuss topics from a position of superior intellect than from one of wisdom and truth. Simplicity and ease are not our concerns here. Our main purpose is to cut through those things that distract us from finding the ultimate truth in anything. Ground zero, the pivot point upon which humanity moves and breathes, is the target we seek. With that firmly in mind, let's talk a bit about capitalism.

---

91 Many economists today believe that wealth-building is the key to solving poverty and even protecting the planet from environmental degradation. In essence, these economists state our planet has resources sufficient for our needs for centuries, if not millennia to come. Also, there is this belief that science and technology will evolve to the point where fewer resources will be used and the ones that are used will be used in a more efficient, clean, and safe manner. Obviously there are a lot of assumptions made in this kind of economic and political thinking. This kind of scenario could develop, but should we base our survival on such assumptions?

The majority of people like to think about, debate, and lay blame upon a system or systems (whether they are democratic, capitalistic, socialistic, communistic, autocratic, or anything else) simply because they believe one or more of them is the cause of many of the world's problems and therefore must be fixed or abandoned. This approach has supplied us with reams of written material, endless discussions on radio and television shows, many a revolution, and heated debates ad infinitum. However, our discussion of capitalism will dispense with the innumerable arguments that address everything but where we're at today, the place and time in which we're in danger of killing the planet and ourselves.

Systems themselves just draw our attention away from the fundamental truths we're after. Yes, it does seem important to debate the pros and cons of one political, economic, or social system over another, but in reality the urge to repair a system, or to reject or replace it, is a complete waste of time, money, and energy if we are unwilling to admit and address the actual source of the problem.

In the United States we have been brought up on the notion that capitalism is the best economic system for the world and humanity. It's pure irony that all of the things that Americans have condemned in other systems have been endemic within our own. Cronyism, nepotism, corruption, greed, and fraud live and breathe just as strongly within capitalistic walls as they do within socialistic walls. This may be a hard truth for the capitalistic cheerleader of business to accept. But let's be honest with ourselves: no one economic system has the monopoly on fairness or justice for all.

Capitalism, socialism, and the variety of other systems are a marvelous case study of how we think and act on an everyday level. Like media's constant bombardment of social norms, all government and economic systems take a similar path of self-promotion and crowd control (i.e., the handling of how the population thinks and acts). Most people have little understanding of how their particular government or economic tradition has influenced their thoughts and actions and conditioned their subconscious minds to accept or reject fact that in many cases is more fiction than truth. Keep this one fact in mind as we move forward, whoever controls the overall narrative controls the subconscious minds of millions.

Winston Churchill once stated, "Capitalism is the worst of all economic systems, except for all the others." That statement has truth to it. Beyond a doubt, capitalism seems to be the main driver behind the eradication of poverty. One only has to look at humanity's short history to realize how commerce increases technological innovation and national wealth, and, in a general sense, allows people to rise above impoverished living conditions. Indeed, this is capitalism's shining star, an ideal worth striving for. Should not this ideal be kept and continually supported?[92]

Any cursory study of human history will reveal that humans lived a hand-to-mouth existence for millennia until invention and trade became established as ways to escape poverty. Even with the socialistic systems we have today, trade and productive power are quintessential ideas that make the eradication of poverty possible. These elements have been removed from some socialistic systems of the past, but if one looks close enough, those particular systems never lasted in the wider context of raising the population above subsistence living. For the most part, capitalistic systems allow people the opportunity to earn higher incomes, to escape poverty, and to grease the wheels for further innovation which eventually spreads the material gains throughout society.[93] This is not a theory but a proven fact. And yet, capitalism has also exploited the poor of many nations and kept them in bondage to the capitalistic machine; again, not a theory but fact. Does one statement contradict the other? Absolutely. Does one statement negate the other?

---

92 The concept captured in the word *ideal* is paramount to a better life, i.e., striving for and choosing a path for altruistic reasons. How many times have you heard the comment that ideals are for the naive and ignorant? In a Machiavellian world where politics, money, and pragmatism rule the day, believing in, much less striving for, a moral ideal is tantamount to proclaiming one's idiocy to those who know better. How this sophisticated and intelligent group of people missed learning the history of, the importance of, and the objective reality behind the concepts of ideals, moral values, ethics, and justice is beyond comprehension. One really does wonder who is pursuing foolishness and who understands the universality of that which makes life worth living?

93 This last statement is the ideal scenario, but is not always the case.

Absolutely not. As you may recall from the preface, logical contradictions do exist and are still valid regardless of our notions on logic and how it should be applied. Also, if you recall the way that linear thinking shapes our ideas of logic and reality, you will see no contradiction between the negatives and the positives of a capitalistic system; they are one and the same.

Is capitalism bad for society? The answer is an unequivocal no. But does that mean the system cannot be improved upon in order to reduce the bad elements in favor of the good? That, my friend, is always the question in life, and you intuitively know the answer. The universal truth has always been to seek the higher ground. Does capitalism have bad elements that can be improved upon? Yes. Does socialism have bad elements that can be made better? Of course. Is one system better than the other? No. Can socialism and capitalism actually complement each other? Absolutely, and they should complement one another—if we are looking at things in their true light. The notion that capitalism (in its current form) is the best possible solution for society's health and well-being is misguided at best. The absolute truth of the matter is that we need an economic system that is not explicitly exclusive of things that support the universal truths that give life its ultimate meaning.

Now let's return to a question asked earlier: Is there any justice or moral high ground in keeping certain institutions and businesses in power, to continually reap ever more wealth and dominance for themselves at the expense of humanity and the planet?

Reflect with me for a moment on the social systems we have today, the result of maintaining certain institutions and businesses in which the "smart" and "intelligent" pull the levers of business and government.

- We have a social system that fosters wealth accumulation over poverty eradication.[94] A social system that allows (as of this

---

94 This does not necessarily mean that capitalism is bad in and of itself. However, in its current form, capitalism clearly contributes to the destruction of the planet and of people in many ways and on many levels.

writing) more than five billion people to live in poverty. This population lives without health care, the means to feed a family, and in most cases without an avenue to escape their poverty. The poor of the world literally fight to survive, while the remaining 20 percent of the population concerns itself with health, youthful looks, and the latest forms of entertainment. Our current social system squanders the resources, time, and capital that could be used to save a child, a population, or even the earth.

- We also have a social system that bombards us with images of dysfunctional behaviors that promote self-indulgence, sarcasm, ridicule of others, promiscuous sexual practices, lying, and violence as normal and acceptable forms of behavior.

Acceptance of these perceived societal norms has contributed to and magnified drug and alcohol additions, the collapse of the nuclear family, sexual abuse and disease, and plenty of other ills. The more common the thought or act becomes, the more it self-perpetuates and self-justifies itself as an acceptable or absolute reality. Just one example among legions would be the AIDS epidemic. AIDS has killed millions of innocent people and has destroyed lives and families on an incalculable scale. AIDS is primarily transmitted through sexual contact, and the new reality's endorsement of reckless sexual practices only fuels the pandemic. When untruths become the norms in society, such as granting people the license or the right to have sex whenever and with whomever they want, then things like AIDS and the human papillomavirus (HPV) become everyday problems. This was not always the case.

We know this new kind of reality and truth are subjective because of what we already know from human history, i.e. the dysfunctions we deal with today are not absolute realities. Those dysfunctions are a legacy of our carte-blanche acceptance of Enlightenment beliefs (unrestrained freedom without accountability). Epidemic levels of sexually transmitted diseases are not an inevitable reality of life. Once again, do we define universal truths, or do those universal truths define us? History can tell us a lot about what is objective and what is subjective truth. With that said, let's briefly discuss how history is used in revealing truth and reality.

There are some—in fact quite a few—who believe that history is an invention at best and illusion at worst. If you recall our discussions on truth, you will see a few parallels with the work of sifting historical fallacy from fact. While it is true that history can be made by those who write it, it is also true that the overall landscape of history choreographs facts. Let's talk about a few of them.

While we may not know the finer details of the Black Death which killed millions in the fourteenth century, we do know that it happened and that it was a factual event. Likewise, we may not be able to accurately measure the amount of rainfall that fell in southern Utah seven hundred years ago, but we can use data from tree rings and geologic core samples to determine that rain did indeed fall and in roughly what amounts it fell. Along that same line of thought, we may not know what the southwest Indians thought about nine hundred years ago, but we do know what they made, the food they ate, how they traveled and traded their goods, and a whole host of other facts about their lives.

History reveals certain and undeniable truths. Don't throw the baby out with the bathwater just because some writers of history have opinions and biases. Remember that deductive reasoning is the key! Compare evidence from many different sources (convergent evidence) to validate fact and discount fiction. Just as we have grey areas concerning our search for truth, we also have grey areas where history is concerned. It is those grey areas that must be eliminated if we are to arrive at the objective truth of the event, time, or narrative. If one looks at history through the eyes of one, two, or even several historians, then our vision is blurred. If we look at history through many different mediums, such as archeology, geology, linguistics, climatology, agriculture, and biology, the blurred vision becomes clear.

The arrow of time itself is as axiomatic as life and death. It is nothing less than history in motion! Hence, history is reality, a unified truth within our closed system. History is not made by wars, winners or losers, political movements, or even scientific discoveries. No one thing makes history; history is an amalgam of all. It is this amalgam that reveals the truths history has to tell. As long as we concentrate on the tree instead of the forest, history will always present opinions instead of truths.

So, when our discussions refer to history, they will refer to the unified whole. It is always the greater sum of the individual parts that gives us access to truth. After all, why should we be underwhelmed by the parts when we can be overwhelmed by the whole?

The smart and intelligent people who think and act using Style II and III methodologies of reasoning (using a lot of compartmentalization in reasoning) have enforced and promulgated the idea that it's morally acceptable to think and act as self-focused free agents (an adolescent perspective, to say the least) regardless of how those actions may affect the lives of others.

At one time in history, being wealthy was the biggest risk factor for extreme narcissism. Today, however, sufficient exposure to westernized cultures and media seems to be enough. Consequently, so long as Western values of the supremacy of individual rights are repeatedly expressed and taught, the end results will always be narcissistic and destructive in nature. This is not to say that individuals who have avoided contact with westernized cultures (something practically unheard of today) do not suffer from narcissistic habits; instead, I wish to point out that westernized cultures have actively pursued systems that intensify and market these values, and ultimately indoctrinate society with self-centered ideologies. It is fair to say that the world today is adopting, or in many cases is being forced to adopt (via trade agreements, financial arrangement, United Nations sanctions, and the like) westernized systems that ultimately culminate in the acceptance of westernized values.[95]

Two respected and well-known social scientists, Solomon Asch and Stanley Milgram, systematically proved that the majority of people (Style II and III thinkers) conform to group dynamics or perceived authority figures in spite of and in violation of traditional moral and ethical values (i.e., those objective truths that sustain absolute reality). Members of a group or organization will ignore or suppress objective truth(s) in an effort to maintain the group's own subjective realities. In essence, Style II and III thinkers are easily manipulated and are routinely used as agents of distortion and falsehood while justifying themselves as loyal members of a group, society, or organization. There are many non-intuitive layers to this last statement; let's examine a few of them.

There are countless instances of businesses' motivation for greater and greater profits causing individuals within those businesses to act in unethical or criminal ways. These include lying to the public, creating inferior products, using deceptive accounting methods, bribing public officials, market manipulation, producing products with health risk, using false advertising, and condoning closed-door meetings and undisclosed transactions that eventually force products and services upon the public without its input or consent (the U.S. food system alone contains plastics, genetically modified plants and animals, synthetic hormones, antibiotics, heavy metals, and toxic chemicals, to name only a few, all being used in the public food system without the public's consent). These actions are all taken in the name of freedom, the freedom for businesses to be unencumbered by governments, the public, and the objective truths that define and support life.[96] The question needs to be asked, is this kind of freedom really free?

The whole concept of freedom hinges upon a society's understanding of responsibility. When a society decouples freedom from responsibility,

---

95 Westernized media, trade agreements, and westernized economies are the main tools that indoctrinate the world of today.

96 This is one of many examples of capitalism in conflict with democratic principles.

life slowly degenerates into chaos.[97] We are witnessing this unrestrained and irresponsible use of freedom not only in the business world, but in our private lives as well. Freedom is never free when it disrupts the health and well-being of other people or the planet upon which we depend. Ironically for westernized cultures, unrestrained freedom of this type translates into a host of physical pathologies: cancers, immune deficiencies, sexual disorders, sexually transmitted diseases, cardiovascular diseases, chronic neurological and respiratory disorders, and so on, all consequences of what we allow into our bodies and minds. Is this really freedom at its progressive best? Not only does 80 percent of the world live in poverty,[98] and not only is the earth being systematically destroyed, but our so-called free and enlightened society is destroying itself with those same definitions of *freedom* (i.e. progressive regression). Somewhere along the way, modernity has severed responsibility from our views on freedom, and because of this, we are creating a new reality that is not only non-reality based, but counterproductive to life.

I have heard the question asked many times: *What can I do against all the injustices of the world? I'm only one person, and it would take an army of people to make a difference.* This is classic Style II and III thinking (that is, "I'm not going to change if others don't. If I drastically change my way of doing things, I will have to live a different lifestyle than my peers, family, or the culture in which I find myself."). As long as Style II and III thinkers are afraid of what others think, or of living a life of sacrifice in order to make another's more bearable, then the status quo will never change. Is it any wonder that the have-nots of the world feel a growing resentment of those who live lives of ease and luxury? Let's walk a bit further down this road.

---

97 If you remember our discussion on quantum thinking, you should understand the reality and unity of the two words *freedom* and *responsibility*. You cannot separate one from the other without rendering both meaningless. They are one in essence and cannot be made otherwise.

98 Living on five dollars a day or less.

In the Americas, Style II and III thinkers turned a blind eye to the plight of the Indians (a situation that European and American society created). From as early as the 1500's through the end of the nineteenth century, the majority of the population sanctioned murder, rape, larceny, displacement, and enslavement of American Indians (shameful, to say the least, and criminal in its reality). During the eighteenth and nineteenth centuries westerners enslaved Africans, adding more than two hundred years' worth of discrimination. The unrestricted exploitation and outright brutal treatment of nonwhite Americans persisted for at least eight generations; twenty-five U.S. presidents came and went before any semblance of justice was done. Were there any who opposed these injustices? Indeed there were. But then, as now, they were outnumbered by the majority (ostensibly Style II and III thinkers), who had fixed their eyes on the status quo and wealth accumulation—a majority I might add that were backed by the U.S. government and the highest court in the land.[99]

So once again, why bring history to the forefront of our discussions? After all, these things happened in the past, and we as a people are beyond such things in modern times. We're progressive, remember? As a people, however, we are no more advanced or progressive than our ancestors if we continue to live in a closed and illusionary world (where discrimination is out of sight, out of mind, and therefore none of my concern). I'm quite sure the citizens of Boston, New York City, and Philadelphia in the 1800s acted according to their times. For example, in the eighteenth and nineteenth centuries, Indian Territory was considered a foreign land, a place far removed from progressive society. Indian Territory was thought to be inhabited by impoverished and unintelligent Indians and thus ripe for exploitation by a more advanced society. Nothing has changed: then as now, the poor are manipulated by the elitists of the world. Where once we used coercion, bribery, and false promises to achieve our goals of wealth and power, we now use more sophisticated and refined methods of coercion,

---

99 See, for example, the Supreme Court decisions of Cherokee Nation vs. State of Georgia (1831) and Dred Scott vs. Sanford (1857).

130

bribery, and false promises to achieve wealth and power within the wider world. If you doubt this last statement, spend a little time examining today's financial systems and how they're designed to exploit the poor while enriching the coffers of the wealthy (i.e., fair wages denied to the poor, corporate rights displacing human rights, free-market policies supplanting regulatory and/or public policies).[100]

History provides a good base from which to measure. In the examples given above we can measure our progress, or lack thereof, from historical to contemporary times. History gives us a baseline upon which to judge and offers lessons to learn from past experiences. These historical examples emphasize universal truths that eliminate many of our modern assumptions and biases. Considering our examples above, the obvious deduction would call into question the very notion of a democracy in which the majority of people (Style II and III thinkers) are willing to turn a blind eye to notions of liberty and justice for all. How long will it take before Style II and III thinkers face objective reality? Because of our moral ignorance or sheer obstinacy, millions are denied happiness, well-being, and life. The moral price is always the greatest among the innocent. Because of people like Thomas Jefferson, Andrew Jackson, Robert E. Lee, and thousands of others (all of whom lacked the moral fiber to do the right thing), millions of innocents paid with their lives, and many more millions carry the scars and baggage that is passed from generation to generation even unto the present day.[101]

---

100 For interested readers, the following examples are well worth considering: the demise of the Glass-Steagall Act; the Supreme Court decision in Citizens United vs. Federal Elections Commission; the power of the World Trade Organization (WTO) over human and sovereign rights and the exclusion of public oversight and participation from it; members' rights under the General Agreement on Trade in Services (GATS); the privatization of government during the Thatcher/Reagan years; the World Bank's role in lending money to impoverished and developing nations and the restrictions placed on such lending by the International Monetary Fund (IMF); the stances of the World Bank, the IMF, and the United Nations (UN) on offering aid to poor countries; and the exclusions and focus points of business pricing models.

101 This last sentence should hit an emotional nerve with most Americans. If it did with you, just remember and reflect on our discussions so far regarding Style II and III thinkers and their methods of creating and justifying their own realities.

THE MATRIX OF TRUTH

Because Style II and III thinkers are afraid to take a stand against the status quo (whatever that might be in each person's life) and to step outside of their comfort zones, innumerable souls are placed on trajectories leading to disharmony, decay, and early death. Everyone pays for such moral failures. Not a single person escapes the dysfunctional cancer that haunts us today. This extreme dysfunction is plain to see if we but open our eyes to see it. How do we view people from outside our accepted groups? How and what do we spend our time and money on? What do we carry as emotional baggage, and what are our attitudes concerning pride and humility?

The examples given above are but a few of the thousands of ways that Style II and III thinkers are easily manipulated and readily compromise or forfeit objective truths in their allegiance to systems based in subjective realities. Lest you exclude yourself from blindly following the status quo and committing such horrors, bear in mind that many social scientists since the Asch and Milgram experiments have repeatedly observed similar thinking and behaviors in each successive generation, up to and including our own. Do any of these examples tell us anything about our ideas and positions concerning benevolence, integrity, and value (remember our discussion on compartmentalization)? Does it tell us anything about our proximity to those who suffer, whether they are our neighbors next door, across the city, or in a foreign country, while we spend time and money eating, drinking, and making merry? Do these examples say anything about the systems and the subjective realities we have created, maintained, and constantly accept as foregone conclusions? As the world slowly spirals towards its death, do we take a stand for what's right or do we sit back and go with the flow? Can you and I disassociate ourselves from the status quo enough to see reality for what it is and not what we have accepted it to be? If you count yourself blessed to live in a wealthy nation, ask yourself, what do you strive for in life? If humanity is to save itself, it will be by changing the thinking and behaviors of the vast majority of people (Style II and III thinkers).

Anything less, and we might as well say good-bye to everything we hold dear.[102]

Many individuals will no doubt become indignant after reading the less-than-flattering description of Style II and II thinkers. There is a very real possibility that you yourself are a Style II or III thinker. If you are in this indignant frame of mind, I want to assure you of a few things before we continue. First, even though I'd like to believe I'm not a Style II or III thinker myself, when I sit back and assess many of my desires and actions, there is no doubt that I spend much of my time in that very domain. So if I come off as sanctimonious, critical, or condemning of Style II and III thinkers, please realize that I reprove myself as much as anyone. We all have affinities, no matter our thinking style, with Style II and III thinkers. We can all see a part of ourselves using Style II and III thinking. The point to remember is, what do we do with that information? That choice, of course, is entirely up to you. My hope is that we all strive to seek and live true reality in our lives.

Also, my intentions are never to demean or discourage the human spirit. It's an established fact that no matter who we are, we each have the ability to rise above the definitions and labels that others place upon us, mine included. We need only view our own history to realize that we have the ability to overcome our categorizations.[103] Second, as you will soon discover, all of us (no matter our thinking style) have a great deal of work to do in order to achieve a very basic and life-altering goal, that of saving ourselves from a multitude of abnormalities and saving the planet from the daily destruction we inflict upon it. All of us, not just one thinking style or another, are guilty of being bystanders or an occasional compartmental-ized participant when we should be full-time players.

---

102 None wish to think about the road we currently travel, but nonetheless, we cannot deny reality. The nexus upon which everything now rests is in the hands of Style II and III thinkers. The quintessential question of our time is, can that majority rise up and change the course of history?

103 See, for example, the abolition of slavery (early to mid-1800s), Mahatma Gandhi's movement for an independent India (1947), the Hungarian Revolution (1956), the civil rights acts of the 1960s and '70s, and the general strike that took place in France (1968).

Auditing life by immersing ourselves in self-seeking behaviors is counterproductive to our growth as individuals and to our society as a whole. But time is too short for us to sit passively on the sidelines while our planet shudders in its final death throes. We all need to change. The subjective realities we build for ourselves are on a collision course with the supreme reality. We have dodged, abused, and covered up the universal truths that give our planet life, but in the end, we cannot escape what ultimate reality has in store for us. It will be destruction and death if we don't change. True reality always has the last word. It's time we wake up to this fact and hope and pray we're not too late.[104]

---

104 Remember, humanity never defines true reality. True reality always defines us, and it will ultimately have the last word.

- PERCEPTIONS, FREEWILL, AND ATTITUDE -

"We who lived in concentration camps can remember the men who walked through the huts comforting others, giving away their last piece of bread. They may have been few in number, but they offer sufficient proof that everything can be taken from a man but one thing; the last of the human freedoms- to choose one's attitude in any given set of circumstances."

-Viktor E. Frankl

When I was born, John F. Kennedy was the president of the United States, there were still Nazi war criminals loose in the world, television was growing in popularity, and the Russians were building the Berlin Wall. Struggles would soon follow for the civil rights of black Americans, as well as the complications of the Vietnam War, changes in sexual mores, a new environmental awakening, a growing drug culture, and a shift from traditional values to laissez-faire ways of thinking and acting were taking place.

As a boy growing up during this time, I was oblivious to the trans-formations and the new norms that were taking the Western world by storm. For me and many of my friends it was a magic time to be alive. We could dress how we wished, grow long hair and beards if we liked, smoke marijuana, have sex whenever we wished, tell authority figures to shove it, and pretty much run our lives as we saw fit without feeling guilty or condemned. It was complete freedom, and it was intoxicating, liberating—independence at its best. For the new generation, those youths growing up in the 1960s and 1970s, it was a great time to be alive! We were experiencing the birth of the greatest social revolution in the history of humanity. And we believed it would completely transform and change society for the better.

By the end of the 1970s, I had noticed that the revolutionary magic was morphing into something else. By the 1980s, although I could not put a finger on the causes, I sensed that a great war was being fought. Many of the battles seemed to take place in the shadows. Some battles were visible to all, like those of the civil rights movement and environmental policy, but other battles, those behind the scenes, seemed to injure, wound, and mutilate combatants on a much more personal level. Eventually awareness came to me. The world had not evolved as we had hoped for. The beautiful canvas we had originally painted had devolved into the disfigured portrait of Dorian Gray. The changes were so subtle that each individual was deceived in remembering its former beauty, even as the portrait gradually moved from something beautiful to something grotesque and unrecognizable.

Indeed, the world had changed. What once was light had become dark. Innocence had turned to despair and hardness, simple laughter had become tainted sarcasm, and pragmatism replaced altruism. At one time, society looked for beauty beyond itself and reached for ideals that were noble. Much of that time was now lost. The air we breathed was being replaced with carbon monoxide, arsenic, and mercury. The food we ate was now laced with drugs, chemicals, and modified substances. The water we drank carried solvents, fuels, and fertilizers. If that was not bad enough, love in the home was replaced with a political agenda: i.e. *What can I get from the situation* which is ultimately not good for the health and/or the security of the family. Physical pleasure was corrupted into overeating, drug use, hedonism, pornography, and the myriad of other things with which people seek to amuse themselves with. Hence, no one seemed to know what innocence meant anymore.

When one grows up in a specific culture (be it that of the ancient Greeks, the Sioux Indians, or contemporary life in the United States), one thinks and acts according to the cultural norms of that particular society. Everyone grows up with those norms, learns them, and believes them to be *agreed-upon* values and truths. Now, as the world becomes more connected through technology, cultural norms that were once isolated to a specific region are being conveyed on a global scale.[105]

As a young man, I accepted my culture's norms as immutable facts of life. I had been indoctrinated into these cultural norms from birth, and they—and thus I—bore society's stamp of approval. In many ways, the legacy of the cultural revolution of the 1960s and 1970s is the foundation of today's norms. These norms take many forms and occupy many levels of our subconscious and conscious minds. Until we can identify cultural norms that are destructive to truth, we can never hope to go beyond our-selves, much less hope to draw closer to the universals that give us life. As you've probably guessed, we are going to take a look behind the curtain of what's considered "normal," "correct," and "acceptable" within our soci-ety in order to strip the illusions away from the truth.

Personal Perceptions

Perceptions are the direct result of what we hold in our minds. Absolutely nothing can quantify the mental constructs held by each individual. Because of this, we must look beyond quantifiable systems to other sources of knowl-edge and reasoning.

Every human being who has ever lived, and every person who has yet to live, is at a severe disadvantage in discovering the reality of anything if they don't acknowledge how their brains develop, structure information, and then perceive the world around them. Every person's brain houses a finely constructed system, one that is constructed and structured as new information is received. When we are infants, the brain has little structure

---

105 The more obvious cultural norms that have spread worldwide are those associated with westernized cultures obsessed with entertainment in music, games, movies, shows, sports, and so on, and the not-so-obvious norms doled out and expressed through the principles of capitalism. Capitalism has become a worldwide cultural norm through free-trade agreements, international banking standards, the operations of multinational corporations, and even mass media's glamorization of wealth accumulation as life's primary goal.

upon which to base decisions. The infant has no concept of heat until he or she burns themselves on something that is truly hot. After that experience, the infant builds a mental construct (a structure) of the event. As time goes on and the infant experiences more things, more mental constructs are added to the existing structure, and the mind slowly builds a perceptual dwelling in which to live. There is utility in the process—that being, it keeps the mind from having to start from scratch every time it receives new information. As we mature, the mind takes new input (information, the mind's raw material) and tries to fit it into the framework of our perceptual house. If the information seems to fit within our perceptual design, we add it to our current mental structure. If the information does not fit, however, the mind usually does one of two things: it rejects the information out of hand as an unrecognizable piece of building material, or it recognizes it as potentially useful and stores it for future reference.

This is where things get a bit out of hand and why there are so many definitions of reality and truth today. In essence, the mind sees and comprehends only what it wants to comprehend, depending on the perceptual structure that is in place and maintained by each person. Hence, if a person has used the building materials of, say, Euclidean geometry to construct their perceptual house, then any non-Euclidean building material would be rejected as nonsense. When much or all of the existing structure would have to be torn down and rebuilt to accommodate the new and foreign input, most people are loath to do the work.

If you are a mathematician or a scientist, or if you just have a passion for these fields, you probably realized why I used non-Euclidean geometry as an example. We have already discussed Einstein's use of a cosmological constant. Einstein's perceptual dwelling seemed rock-solid, but his theory of relativity could not be used within it, or indeed within the mental structures prevalent at the time. So, instead of tearing down all or part of the scientific community's structural edifice, Einstein sought a building material that would act as an intermediary, a bridge between the two incompatible materials. Remember, people dislike tearing down existing structures, especially the ones they live in, maintain, and have nurtured for many years! Individuals put a lot of time and effort into building their mental constructs, tearing them down (in part or in whole) is a threat to their very being. People dislike retreating, regardless of where the road leads.

In the mid-nineteenth century, Euclidean geometry was the perceptual structure upon which the universe operated. It was not just accepted, but "proved." Ever since Euclidean geometry was established as a perceptual point of reality, any proposition offering a differing point of view had been considered absurd. This, of course, begs the question of how our perceptions have been incorporated into our mental structures of reality today. To answer that, and to get our intellectual juices flowing, take a moment to think about the following perceptual issues:

1. We view reality as it is recognized, accepted, and incorporated by the structure of the mind (usually in a linear and heuristic fashion). The structure of your brain only accepts a reality that closely matches its current structure. Your mind places certain expectations on the information it receives. If it did otherwise, the information would make little sense, i.e. we would be starting from scratch every time a new piece of information was presented.

2. Science and mathematics are all built upon constructs. Both science and mathematics build equations, instruments, and theories to fit within a desired mental structure. These constructs answer questions that fit within a prebuilt, self-limiting system of thought (i.e., axioms, the scientific model, etc.). Many assumptions made in the mental building process seem true in one environment, but are irrelevant when viewed from another (i.e., an environment other than the scientific one). Throughout history, science has proven to be limited in its pursuit of reality by its self-imposed constraints of the scientific model. (We will address many of these constraints

in discussions to come.)[106] Does this mean we should discount science or mathematics as less-than-worthwhile endeavors? Absolutely not. But bear in mind that science is just one piece of many, one level of several, and one delineation where legions reign. And this is exactly why science and mathematics (based on current methodologies) will never be able to reveal absolute truth and reality by themselves. For that, we will need to combine many fields of study, go back and reexamine our assumptions (usually foundational ones), reassess the mental structures we've built, and learn to think by expanding our linear methodologies while at the same time using nonlinear processes. And yes, we may even have to tear down or demolish some of the mental structures we have so meticulously built over the years, it is a backward-forward step worth taking.

3. A mental structure built on a foundation of knowledge and wisdom extends beyond specific disciplines (as they currently stand) and past traditional styles of thinking. It does not place limits or reject contradictory information out of hand, but rather strives to make truth real in one's personal life (not an individually created or defined truth, but a universal truth) in order to bring us into a right relationship with reality. Instead of being a citizen of the world, it's being a citizen of truth, striving for a culture of truth in all that we do. This kind of mental system is not impossible to construct, but neither is it easy in a world dominated by Enlightened paradigms.

4. When we build a perceptual structure on anything but the truth, it ends up becoming the master by which we are enslaved. This type of

---

106 Richard Feynman once said, "Science is imagination in a straitjacket." Does this sound vaguely familiar? As in life, so goes science: a few individuals seem to get it, the majority do not. Most of the scientific community wears a straitjacket of its own making.

mental structure tells us what we believe and what reality is to be, but it rarely tells us about absolute reality. In turn, the reality we believe determines our worldview, our values, our purpose, and our actions. For the majority of society (specifically Style II and III thinkers), reality is defined by each individual's group or associations. This is why we have so many truths and numerous realities to contend with today. What may be true for one person is erroneous for the next. In addition, this is why we have pragmatic systems (another form of a heuristic system) that adhere to nothing more than what the situation dictates or the individual or group collectively decides. Abortion was once considered murder, today it's an individual's right. Likewise, in the nineteenth century, killing Indians was an individual's right, and today it's considered murder. A military fighter pilot or drone aircraft can kill innocent civilians with few to no consequences, but a soldier on the ground faces court marshal and imprisonment for the same act. Today, if we were to use the pragmatic system of justice, we could not willfully prosecute Hitler or Stalin for mass murder. We could and probably would prosecute both, but it would not be for the sake of justice, given the repeated acquittals U.S. citizens have pronounced for themselves throughout history.[107]

The lesson? There's no sense of justice when truth and reality are reduced to Enlightenment-based, pragmatic models. Pragmatism can always self-justify when no objective values or truths are involved. The only thing

---

107 For example, the United States can attack a country (Vietnam, Cambodia, Grenada, or Iraq) without sufficient cause or threat, killing millions of innocents in the process; there was no justice in these killings, and yet we stand acquitted by our own pragmatic definitions. U.S. justifications for attacking these countries were not unlike the justifications Nazi Germany gave for invading Yugoslavia, Libya, Poland, and the Soviet Union. Another poignant example would be excessive spending by Americans and western Europeans on entertainment while millions die of poverty, disease, and starvation around the world.

pragmatism offers modernity is a justification for acting in one's own self-interest (individually or corporately).

The crux of these implications cannot be overstated. As long as humanity starts from the position that we define reality and not vice versa, there will never be unity or coherence for the objective reality that supports anything singularly and everything wholly. Perceptions built on the foundations of the Enlightenment, specifically those espousing pragmatic values, eventually crumble into chaos and human suffering, exactly the situation we find ourselves in today.

Due to the Enlightenment, in modernity we hold very different ideas about truth and reality than were once held under rationalistic systems of thought. Hence, in modern society, philosophies of pragmatism, materialism, empiricism, and scientific knowledge rule the day.[108] Boiled down to its essential elements, the Enlightenment gave us the idea that our observations and perceptions were the only reality we could ever hope to obtain. If humanity had perceptions only, then we could assume ourselves to be the sole judges of what has occurred (i.e., humans define reality, for there is nothing else to refer to). Taking this argument to the next level, since humans define reality, we must also define our own existence. When we do, the evidence points to a very deterministic process. Humans soon become the *product* of evolution, biology, pleasure, pain, society, heredity, and so on.[109]

---

108 Empiricism is a system of compartmentalized (linearized) knowledge. This is a crucial point to remember as we go forward.

109 Consider Democritus's enlightened reality of deterministic truths and causation. For example, a person thinks he or she has freedom of choice, but in actuality that choice has been determined by some previous event (or choice) that gave him or her the experience and the judgment to make the next choice. That person does nothing but react to stimuli and is only the product of a deterministic universe that reaches back to his or her childhood, birth, to the parents' choices, and ultimately, until the chain of events reaches time alpha (the origin of the arrow of time). Something to bear in mind, from a qualitative and distal perspective, the cosmology of determinism has never been disproved precisely because it is proved daily on an hour by hour, minute by minute perspective.

In many respects, modern people rebel against this deterministic outcome. We either choose to ignore it (believing ourselves to be our own free agents), or we compensate for it by believing that humanity is progressive and will eventually evolve to the point of true freedom. As we continue our discussions, you will come face-to-face with what could be called the "humanistic constant," a constant that modern science and the founders of the Enlightenment tend to overlook, and a neglected point that's causing serious problems today.

Also, because we live in a world shaped by the Enlightenment, most supposed absolutes and universal truths are disregarded as antiquated theories from a previous era.[110] In other words, rationalistic means for discovering reality and truth do not fit modern society's perceptually built structures. As we continue our discussions, we will be addressing the topics of determinism and free will, both of which have significant value to our comprehension of such ideas as meaning, reality, and the truth of anything touching upon our minds.

Empirical systems, by their very nature, are limited by human perception, time, and space. This makes absolute sense in that we live within a closed system (a world of four dimensions, the fourth being time). And because of this, in our quest to make sense of the world, we have built constructs that have practicality and utility within our closed system. These constructs include mathematics, philosophy, science, and human laws.

Hence, in an Enlightened world, humans becomes the final authority on reality and truth. This makes perfect sense to anyone living within a

---

110 A question of logic? The philosopher Lucretius once wrote, "You cannot definitively say there are no absolutes, for by pronouncing it so, you have created your own." Absolutes surround and support us. Food is an absolute for human beings, for without it we die; the same can be said for water and oxygen. The energy that holds together the subatomic world is an absolute. Children who are neglected or abused in their formative years will always have some form (if not multiple forms) of dysfunctional thoughts and/or behaviors as adults. The list of absolutes is endless, and yet most of us don't see the forest (i.e., the absolutes) for the trees (the relative). Everything that seems to be of a relative or subjective nature has its origins in an absolute. To hold the view that life is relative is to deny the truth of the reality that surrounds us.

closed system, because in that system, reality and truth do not exist without a human being present to experience and define them. And since we have created and or discovered structures that have utility within our closed system, humanity quite naturally concludes that these structures and constructs are empirical and objective truths in and of themselves. Therefore, for the vast majority of people, an Enlightened reality is the only reality and truth there is! Anything beyond our scientific and Enlightenment-shaped claims to truth would be nothing but nonsensical and irrelevant theories at best.

I suppose if Einstein, Planck, Schrödinger, and others had not come along, we would never have had a scientific basis from which to question the Enlightenment model of reality. It is interesting to observe, in the history of humanity, how we return to and recycle ideas from previous ages. (Historians know this fact well). As with the advances of Copernicus, Newton, and Darwin, there is almost always a time lag between the discoveries themselves and a time when the popular imagination understands the full impact of those discoveries. Even today, the scientific community has not fully comprehended the far-reaching implications of relativity and quantum mechanics and the paradigm shifts they hold for empirical belief systems.

Today, we are caught in just such a time lag, a cycle of returning to ideas we once thought had been put to bed (i.e., the enormous implications and unresolved mysteries surrounding relativity, quantum mechanics, and the ever-expanding fields of cosmology). These advances in knowledge have pried open a door once thought to be shut, locked, and forever forgotten. And yes, you may have guessed it, this particular door has *rationalism* carved upon it. In the final analysis, if we choose to seriously consider the implications of the latest scientific and epistemological advancements, we'll need to reopen that door to some rationalistic concepts. Much of the knowledge and evidence locked behind this door is badly needed in order to open up new avenues of understanding and to seek answers for seemingly unanswerable questions. We need an interdisciplinary research approach that analyzes the old and new, that reexamines the seen (e.g., the brain, physical matter, linearized conceptions) and the unseen (e.g., the mind, dark matter, quantum perceptions), and a synthesis of the closed microworld in which we live and breathe with the macroworld in which we happen to occupy a very insignificant place.

We have passed through several revolutions of thought since we first began to record our ideas via the written word. From the Archaic Era to the Reformation, we accepted the idea that reality formed an absolute and single system. This single system of knowledge was usually tied to a mystical or supernatural element (e.g., a god or gods).

As humans moved into the Renaissance and then the Enlightenment eras, we seemed less tied to the land, primarily due to new technologies. More of us were increasingly able to attain levels of education equal to or greater than that of the nobles and clergy.[111] As an educated middle class developed, the church found that its monopoly on knowledge and power was waning. Eventually, this burgeoning educated class gave rise to the age of reason, when average people began to think for themselves, rather than letting authority figures think for them. The common folk began to make their own rules and decided what their futures should be. From that time until today, we have basically defined and created our own reality via institutions, laws, business practices, and education, in thought as well as in deed. In fact, by the late 1800s, with the advances in chemistry, medicine, physics, biology, and other fields of study, many had thought that humanity had almost reached its zenith of knowledge and discovery. In 1894, world-renowned physicist Albert Michelson stated that "Most of the grand and underlying principles of reality have been firmly established." When Sir Isaac Newton solved the problems of motion, space, and time, everything seemed to fall into place. Between mathematics and science, the universe became a very determined and predictable place. It seemed that we had not only conquered nature, but were also on the brink of

---

111 I am purposely keeping the discussion focused on European history, since the majority of our Enlightened ideas were both born and nurtured in Europe.

solving the most complex mysteries of life (sound familiar, we still believe this today). The primary mechanism for this certitude was the exponential growth of knowledge via the physical sciences. Ironically and unwittingly, science has now opened a new door—or maybe, more appropriately, redis-covered an old one—that has called our Enlightened reality into question. Believe it or not, a new storm is developing on the horizon, one that will usher in a new renaissance in our understanding of the nature of truth and reality, not by way of empiricism, but by way of rationalism!

Science continually discovers more evidence for a transcendent sys-tem, one that is slowly unraveling the system we have come to depend on. Most have not grasped the implications or consequences of the impending shift. Inflationary cosmology, quantum fields, dark energy, dark matter, the cosmological constant, simultaneity, parity inversion, entangled states, and other concepts all point to a transcendent reality that stands apart from our closed experiential/empirical system. You may not realize it, but these scientific findings are counterintuitive to the Enlightenment model that has been codified and embraced for the past two hundred years. But you are not alone; even many in the scientific community have not caught the full meaning or significance of the reality that stands beyond Berkeley's observations, the anthropic principle, or even the dilemma of Schrödinger's cat.[112] In spite of the obstacles, we now know that the reality we observe in our Enlightened world has very little to do with the transcendent real-ity that lies beyond it. Indeed, a few people are now coming to terms with the epistemological realization that *We know that we don't know* and are conscious of the fact that reality is much more than that which is bounded by empiricist ideals. In the end it would appear that those things we once viewed as science fiction might in fact be very natural parts of a dimen-sionless truth or an absolute reality.

---

112 This is reminiscent of Einstein's rewriting of the Newtonian laws that had been so firmly entrenched in the scientific community . Newton's ideas were already, in the minds of the scientific community, established as fact. Einstein's theories just didn't fit within the established framework of Newtonian physics.

I would like to leave you with one final thought. Enlightenment-born systems start from the position that our senses are all we have to work with. It goes without saying that this position is subjective at best. Humans are faced with overwhelming variables and almost impenetrable complexities that defy our linear, reductionist, heuristic methods of reasoning. Because of these complexities, we assume that reality and truth exist only within a linear world, a closed system. For example, science today views general relativity as being true on one level and quantum mechanics as being true on another, but both become inconsistent when combined and viewed on the same level (i.e., to produce a predictive state $\emptyset$). [113] When physicists try to combine quantum mechanics and general relativity, via mathematics, the result is always the same: infinity. Interestingly enough, in the field of physics, when an equation or experiment yields an infinite answer, it's automatically suspect and viewed as a nonsensical product, essentially a junk answer. The reason for this is that we (and not just scientists) operate from the paradigm of linearity. For instance, infinity can't be quantified: we have no direct experience (Berkeley) with infinity, and our minds cannot grasp a concept that has no beginning or ending. We'd like to say our understanding of the concept is broad, but that's wishful thinking at best. A linear approach will never fully comprehend something that's immeasurable and has no starting or ending points.

To highlight this even further, let's focus for a moment on Albert Einstein. When Einstein created his equations for the theory of general relativity, he intuitively saw the implications they would have on the so-called "known" characteristics of the universe. His equations suggested a universe that flew in the face of reason and established facts—a universe

---

113 General relativity describes very accurately the macroworld of stars, planets, and galaxies. Quantum mechanics describes very accurately the microworld of molecules, atoms, and subatomic particles. Both theories appear to be true in their own domains. But how do they work together in order to explain reality as a whole? Hence, we operate as if there are two separate truths or realities when in fact we know that the two systems must be connected in order to make the universe operate as it does.

that was still expanding, for instance.[114] Interestingly enough, for many years even Einstein refused to believe what his own equations seemed to reveal. Because of this, he thought a piece of the puzzle was still missing. After all, everyone (including himself) knew that the universe was unchanging, immutable, and fixed. In the end, and in order to make his equations work with the scientific assumptions of the day, Einstein changed his initial mathematical findings to fit within the scientific views of the twentieth century. By 1917, he had introduced his own cosmological constant to rectify the perceived ambiguity between the results of general relativity and the established "facts" of the science.[115] In light of the future discoveries made by Edwin Hubble and Alexander Friedmann (i.e., the Doppler effect and redshift), we can see that Einstein missed a great opportunity to predict the expansion of the universe. In essence, he was reluctant to abandon what science had thus far established. Needless to say, Einstein's story holds many lessons for us today (the scientific community included)!

Now, let's move forward and discuss an empirical system of belief that has caused much more grief than Einstein's misguided acceptance ever did. In this discussion, we will address empiricist views of free will and determinism.

---

114 A side note about thought experiments: Leibniz came to many of the same conclusions as Einstein without using mathematical equations, and did it two hundred years prior to Einstein's discoveries. Leibniz rejected the established norm of absolute / independent space espoused by Newton. He argued, by deductive reasoning alone, that space had to be inseparable from objects in spatial relations. In other words, space coexisted with everything. Leibniz was ostracized from the scientific community because of his contrary views, and yet he was eventually vindicated (albeit two hundred years after his death).

115 Two points need to be clarified. First, the cosmological constant that Einstein invented in 1917 is not to be confused with the cosmological constant used by physicists today. The two have very different meanings. Second, Einstein's thought experiments, which yielded both the general and special theories of relativity, were the results of a thinking method that synthesized linear and nonlinear complexities. The point is, even with Einstein's ability to think in nonlinear terms, he still succumbed to the perceived linear realities of his day. This, of course, begs the question: What perceived realities do we hold as true that might very well be false?

Free Will

On one level, we tend to believe that we are liberated agents able to make seemingly endless choices whenever and wherever we want. On another level, we freely admit that there are things that make us extremely deterministic (e.g., evolution, genetics, and environment). Are we agents of free will or deterministic robots? Classical logic would say that the answer is either one or the other; nothing can be p and not-p at the same time. We are either free agents (p), or we are determined creatures (not-p).[116]

On one level, the Enlightenment model comes very close to describing our relationship to free will. But in so doing, it neglects to qualify its limits. As we will discover, Enlightenment thinking on free will and free choice leaves out a very important qualifier that actually translates into real-world, real-time lives that are lost or saved. So what is it that the Enlightenment model has failed to incorporate?

A person trained in the use of empiricist ideals is apt to say that the truth of any given situation is completely relative (to the time, place, or persons involved). In other words, two identical situations could be equally true or false, depending on the circumstances involved. So, who is to say that something that appears to be a reality on one level might not be an absurdity on another (e.g., general relativity and quantum mechanics, or the legality of killing your enemy on the battlefield but not in the shopping mall). By using Enlightenment methodologies, truth and reality immediately become circumscribed.

John Dewey was a great advocate for this type of Enlightened reasoning. Dewey's philosophy has evolved into an ethical model in which there is no absolute moral code of right thinking or behavior (i.e., the choices we make and the actions that follow). In essence, the ends (the given practicality of the situation) justify the means (the choice, the behavior,

---

116 This is one example of a contradiction between manmade constructs and systems. Classical logic does not work in every case because it oversimplifies the complex nature of reality.

or the action). Absolute and objective values are sacrificed in order for the individual, group, institution, or nation to define their specific reality as it applies to them in any given situation. The end result is either individualized philosophies or individualized systems of conduct and beliefs. Examples of individualized philosophies would be those espoused by Sartre, Hegel, Dewey, Nietzsche, Marx, Heidegger, Locke, Bentham, and Freud. Examples of individualized systems are: democracy, socialism, communism, capitalism, law, economics, and so forth. Each of these philosophies or systems has legitimized the pragmatic, Enlightened systems of choices that we have in modernity.

The unintended consequence of this approach to free will is the removal of certain and absolute values or truths from the picture. When these absolutes disappear from the human equation, a type of individual and corporate entropy develops. Let's take a moment to understand this concept a little further.

In terms of free will and free choice, we are composed of three parts: environment, genetics, and something called the set choice.

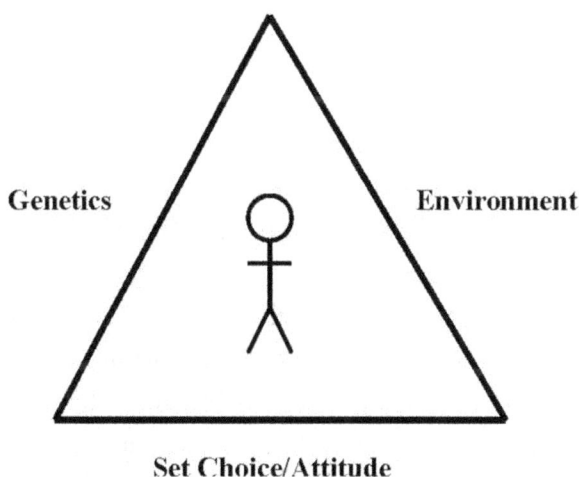

Genetics   Environment

Set Choice/Attitude

The most obvious fact of free will is that at least two-thirds of our existence is predetermined by environment and genetics. In actuality, even more of our free will is compromised when we systematically examine our choices on a day-to-day basis. If you consider a choice and the intended and unintended consequences (or causation) that follow as a result of that choice, you soon realize the linear nature of the choices we continually make. Most of our supposed free-will choices have already been determined by some prior choice, our own, or those of others.

For example, when you drive up to a designated traffic stop, there is a choice or two to make. You can come to a complete stop, look for any oncoming traffic or extenuating conditions, and then proceed accordingly. Or, you could merely slow without stopping, making a few quick glances about before accelerating back to your previous speed. Or, you could choose not to stop or slow down at all and simply ignore the stop, all traffic, pedestrians, and road conditions while charging along at full speed.

The traffic stop examples give the impression that people make a host of free-will choices on a daily basis. But the fact of the matter is quite different. Most of the seemingly free choices we make in life are predicated on some previous choice. In the traffic-stop example, each choice is conscious. If you stopped completely, you made that choice based on some prior choice; perhaps you were previously in an accident because of not stopping, or you were fined by a traffic officer for not stopping. If you slowed but did not stop, you presumably based your choice on a prior choice too: not stopping has never resulted in problems for you. If you did not stop or even slow down, you based that choice on a prior choice, such as feeling the thrill of living dangerously, or the rewards of getting somewhere in a short amount of time. Ninety percent of our choices in life are predetermined by some prior choice or choices. This determined reality has never, and I repeat, *never* been disproved! Because of this fact, we know that our free will is extremely circumscribed.

But does this mean that we are deterministic creatures with all of our seemingly free-will choices predicated on something else? In most cases the answer is a resounding yes! Almost everything about us is determined and predictable when viewed from a historical, cosmological, mathematical, or linear perspective. But when we view our lives through the lens of a person's attitude, we quickly realize that we do indeed have a choice in something. It can be enlightening and maybe a little unsettling to realize that our only real choices in life have to do with attitude.[117]

Attitude

This discussion on attitude will be a bit of a mind-bender for some. Why do you suppose that is? Not only is a relationship between attitude and free will foreign to the Western mind, but it also involves multiple overlapping layers of reality. This way of thinking goes far beyond westernized, Enlightenment-shaped ways of thinking in that it challenges us to think and act in very different ways, ways that we normally do not comprehend on a conscious level.

It is not easy to grasp a truth that dips into quantum ways of thinking while still being presented in a linear fashion. I will attempt to make this truth both linear and quantum at the same time, which usually causes enough confusion to baffle any student to the point of despair. But hang in there! If you can grasp this one universal truth, you will be light-years ahead in understanding your purpose and meaning in life.

Because attitude affects so many things in our lives, it's labeled as the "set choice." In other words, all other choices in life are directly related to and henceforth determined by that one, and only one, set choice. Every

---

117 Attitude and imagination are both created in the mind of each individual. Although these two elements may have links to experience and other outside sources, they can be and often are made without the aid of what we sense or experience in life. And even though imagination (the creative part of the brain) can create new ideas and concepts, it is still subordinate to our attitude(s) towards life and ourselves.

decision we make after our set choice becomes predetermined by the set choice. Another way of saying this is that all of the truly free choices we make in life are founded upon the underlying attitudes we have towards ourselves, others, and our circumstances in life.

The set choice is the one and only master, absolute, and timeless choice we have in life. It is the one area of our lives in which we can exercise absolute freedom. Think back for a moment to the above diagram (the three factors of choice). All of our secondary choices, comprising more than 90 percent of all our choices, can always be traced back to genetics, environment, or attitude. Our set choice however is the only place where our free will is truly exercised and it's our attitude that determines how our free will is revealed. To give this further clarification, examine the drawing and the sliding scale below to see how the concept of attitude and the set choice look in life.

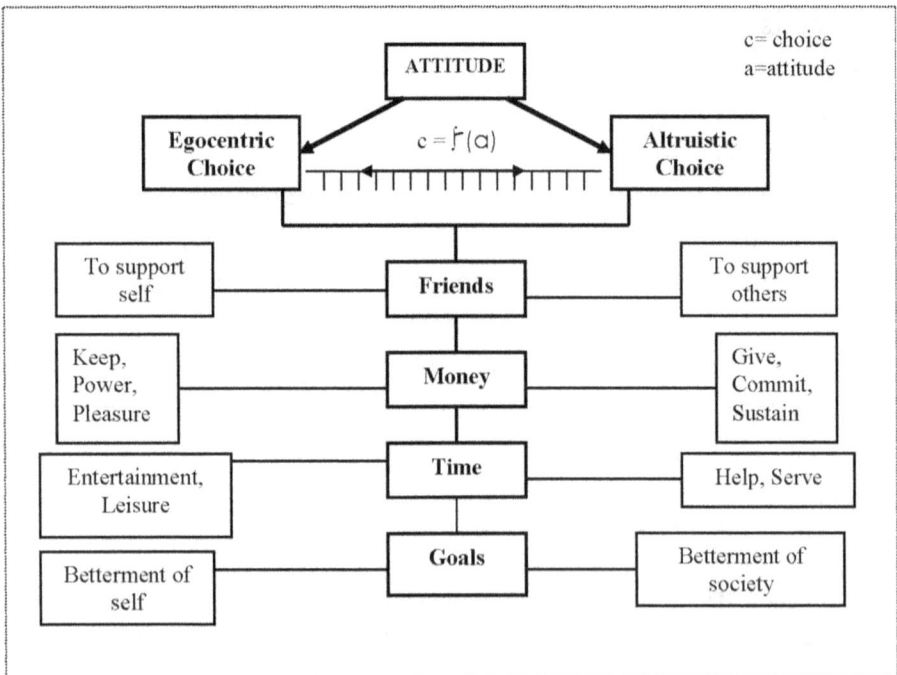

	ATTITUDE			$c =$ choice $a =$ attitude
Egocentric Choice	$c = f(a)$		**Altruistic Choice**	
To support self	**Friends**		To support others	
Keep, Power, Pleasure	**Money**		Give, Commit, Sustain	
Entertainment, Leisure	**Time**		Help, Serve	
Betterment of self	**Goals**		Betterment of society	

154

In the diagram above, you will notice that free choice is always a function of attitude (i.e. $c = f(a)$). Our attitude determines how far to the left or right on the sliding scale between egocentric choices and altruistic choices our free-will choices fall. In other words, each and every secondary decision is weighted toward or biased against self-serving or altruistic attitudes. Once we make our initial set choice, all following choices are subsets of those associative choices with their roots firmly planted in that very first set choice.

It is truly amazing that every day we are presented with one and only one free-will choice, one that is usually presented to us over and over again throughout the day. Think about it. Every day we are presented with choices that are set choices. In other words, there are multiple moments throughout a day in which we can choose to have an attitude that leans towards an altruistic or egocentric world view. Remember that the chain of secondary associative choices can always be broken, at any given moment, by a return to the set choice, that is, making an alternate decision (changing one's attitude) that leans more to self-serving or altruistic ends. A change in attitude equals a change in choice (i.e. $\Delta a = \Delta c$).

Needless to say, the actual truth of free will and free choice is quite different than those espoused by idealism or the Enlightenment. Not only does our set choice involve questions of self and others, it also ultimately means that human life revolves around moral issues, something the Enlightenment aimed to destroy. It's the one thread that unites people with each other, with the world, and with meaning and purpose in life. This truth is nothing less than the atom is to nuclear fission; its meaning for the human race is incredible. Once this bit of knowledge sinks in, you'll see yourself and life in a very different light.

As with just about everything in life, there are multiple connections to things. One of these connections needs to be mentioned before we proceed any further. There's a pernicious fallacy in the Enlightenment-rooted mind that if all the facts are in, then the conclusion is self-evident. This is an insidious fallacy because it appeals to our sense of pride. After all, isn't the scientific method based on obtaining the facts in order to draw the correct conclusions? But for centuries we have missed an essential element in the rubrics of obtaining wisdom: a person can have confidence in the facts and still come to the wrong conclusion about them. Facts and proofs are good only insofar as the intellect is able to fully conceptualize them. To help in understanding this concept, let's contrast it with our contemporary notion of free will, in this example, the paradigm demonstrated by game and choice theory (both of which are progressive models of empiricism and Enlightenment thinking). These are classic examples of having the pragmatic facts, but still missing the expansiveness of absolute truth, reality, and wisdom.

In choice and game theory, economists, sociologists, mathematicians, psychologists, anthropologists, and their peers work on models that attempt to predict choice and how the choosers (the people) will consequently behave. In game theory, there is a classic and well-known test that examines how humans think, decide, and eventually act. This test is called the prisoner's dilemma. Here is how it works.

You are told to imagine that you and a friend have just been arrested for a crime. The police have put you and your friend into separate rooms. A detective walks into your room and tells you that you and your friend are being offered the same deal, and your amount of prison time will depend on your choice between four options. The detective leaves your room at the same time your attorney enters. Now your attorney tells you what the four options are:

1. If you both confess to the crime, each of you will serve two years in prison.

2. If neither of you confesses, and your crime cannot be proven, then both of you will serve three months in jail on lesser charges.

3. If you confess but your friend does not, the police will immediately release you and your friend will serve seven years in prison.

4. If your friend confesses but you do not, the police will immediately release your friend and you will serve seven years in prison.

Your attorney goes on to say, "You must make a choice without talking to or knowing what your friend's choice will be." So, the question becomes, which of the four options are you going to choose?

Today's pragmatic thinking tells you to optimize your choice based solely on what's good for you. In other words, according to game theory, you will optimize your choice for your immediate benefit. Given that assumption, the smartest choice for you would be to confess. If you do confess, at the most you will receive a two-year prison sentence, and at the least you may be immediately freed.

According to game theory, the absolute best choice would be for you and your friend to remain silent. You and your friend would then walk free after serving only three months in prison. Game theory is used to show how seemingly rational choices can have less than optimal or rational outcomes. For example, even though remaining silent is the best outcome for both parties, one is forced to choose a suboptimal choice (i.e. to confess).

Game theory actually shows us that by seeking a purely egocentric outcome, we end up in less than ideal circumstances or situations. On one level, the prisoner's dilemma cautions against basing our decisions on self-interest alone. It's a wonderful example of how set choices can work. However, on another level, the prisoner's dilemma ignores the moral links to justice and deception (i.e., not admitting guilt by remaining silent). By not considering this moral link, game theorists dispense with the more complex issues of individual justice (justice for the victims), individual moral implications (honesty versus deception), and societal ethics (releasing criminals into society). The much deeper lesson to be learned from the prisoner's dilemma is that cooperation alone does not equate with something considered good or even a rational choice based in wisdom. Cooperation that still seeks self-interest before another's rights is still a self-serving decision (an egocentric set choice), and an act of will that will have predictable consequences for other people and for the future.

The prisoner's dilemma, in a greater sense, is a modern-day parable of what idealism, the Enlightenment, and pragmatic thinking have produced. It ignores long-term consequences in favor of short-term outcomes (this sounds very similar to the form of capitalism we practice today - coincidence?). It focuses on the individual instead of the whole, and it makes sweeping assumptions about what is rational, right, and useful while ignoring the moral underpinnings of free will choices. And here we return to our initial concern of assuming we have all the facts which seem to make all the conclusions self-evident. In many cases, Enlightened thinking will claim to have all the facts, and by inference the correct conclusions, but still misses the ultimate truth and reality of the situation altogether.

Before we set aside this discussion of the set choice, we must examine the roles that determinism and causation have in one's life.

Determinism (Causation)

This area of existence is intertwined with and bound into the Matrix of Life, it's intertwined into everything we are and think and do. On a purely human level, objective reality is entangled with deterministic reality.[118] For humanity to even have a glimpse of anything that could be considered

---

118 Deterministic reality is an absolute reality (i.e., the deterministic nature of existence). In an Enlightened world that believes all is relative to person, place, or time, the truths of our determined natures are continually ignored or omitted from discussions having to do with reality and truth.

objective (i.e., reality independent of humanity), we first must accept what cannot be changed by the world, for without this comprehension, our linear minds will be hopelessly lost in a smoke-and-mirrors world in which nothing is truly known.

A deterministic system can focus on causes (i.e., actions) and effects (i.e., results). For example, in a closed system, objective truth is known only by way of effects and results of causes and actions. In other words, objective truth maintains its validity in that its effects are unchanging (i.e., they are repeatable and generate the same results every time). If this sounds a lot like the scientific method, it's because the scientific method relies solely on deterministic principles.

Through deterministic steps, scientists have traced our universe back to its inception, the big bang (BB). We know that gravity exists because it always acts in the same way: it is a determined and repeatable reality. Statistically speaking, social scientists know that if ten children are abused in their formative years, nine of them will have significant social and mental problems in adulthood (cause and effect, actions and results, a determined reality). The overwhelming and quintessential basis for our understanding is that deterministic realities can lead to objective truths. An objective truth always helps in defining objective reality; hence, objective reality is never defined by anything subjective (including human senses). For example, we can define sound and color in many ways (a subjective and personal experience or perception), but that in no way modifies or defines the absolute and deterministic truth behind sound waves or the electromagnetic spectrum.[119] While it is true that we name and attempt to

---

119 Ideas such as sound and color are perceptions of external realities that exist in and of themselves. Enlightenment philosophy has replaced the actual existence of an item with its interpretation. In other words, our minds create the idea of the item and the item therefore cannot exist independently of the mind. But we must always ask ourselves, what is it that we are perceiving? The mind has nothing to interpret if there is nothing to perceive in the first place. The real question is, how do we go from our subjective perceptions to objective reality? There are many ways to accomplish this task, and that's precisely the nature of our quest!

define reality, it is also true that we did not *create* this reality. We can call our discoveries anything we wish, but it does not follow that we created the thing so named. This point is well worth remembering! Now let's dive a little deeper into what this means.

The World of Senses

You may remember that I asked you earlier in our discussion to remember that empiricism and Enlightenment doctrines are based in the belief that humanity can only define reality by way of the five senses. Human senses are electrical impulses that come from the eyes, ears, nose, skin, and tongue. Chemical and electrical impulses convey them to the brain, which in turn processes and conceptualizes those inputs. If sensory inputs were all we had, we could claim nothing more than any other animal species. But the key ingredient—one that the Enlightenment seems to have ignored—is our unique ability to conceptualize things beyond our senses. Mathematics is a prime example of this ability, but it is by no means the only example. Mental discovery can and many times does exist independently of a person's senses. Descartes never realized the full import of his discovery when he established that all thinking presupposes the thinker: "I think, therefore I am." The obvious conclusion drawn from Descartes's discovery is that we can be certain only of our existence and nothing more. But, the deductive ripples on Descartes's pond are the direct consequence of his breakthrough, which affects far more of the world than he knew at the time. In essence, Descartes removed us from the animal kingdom by proving not only that humanity was capable of thought (for animals also think), but also, and far more importantly, that we use our brains for more than processing sensory inputs via a physical means. Descartes and Kant believed that humanity could never know the thing–in–itself. In fact, that is what Descartes believed his discovery was supporting. As the ripples from Descartes's statement spread ever more widely, they affected more things than the initial splash. Think this through and bend your mind past the obvious. What are those things that

Descartes's discovery touch upon? I've already hinted at a few, but there are many more. I reason, I deduce, I refine, and therefore I can know. Can you see the connections that come into play? Can you discover the layers of thought, knowledge, and reason that have been unleashed by that simple yet profound statement that Descartes uttered so long ago?

Human beings are the only animals with the ability to understand the thing-in-itself (this is in direct contrast to Kant's proclamation). Other animals' lives are governed by senses alone, but the truth of human existence lies in the fact that we can postulate beyond ourselves. We're the only animal that can reason past the innate senses of our physical bodies, and the only animal that can create and process thoughts that reach beyond the constraints in which we live.[120] We are self-aware; we know there's more to life than our senses provide. And we are also self-aware in the realization that we can actually learn things beyond Enlightenment models of reality and the unilateral rejections of knowing anything beyond ourselves.[121] If you study the ripples spreading out from Descartes's toss into the intellectual pond, your world will become infinitely larger.

Now let's return to our discussion of attitudes, set choices, and the additional layers that make this particular truth hard to comprehend.

Two Attitudes

In August of 1991, Bob's father called to say that my best friend and brother in heart and soul had been killed. He died while flying a light airplane carrying his aunt and uncle home to New York City. Bob's father told me that the plane's engine had malfunctioned while flying over New Jersey. Apparently, Bob tried to land the plane on a city street when the wings of his aircraft caught on electrical lines and flipped the plane to the pavement

---

120 For example, "We know that we don't know."

121 Given everything that we have discussed to this point, we are faced with a significant question. Think back to the subject of free will and the set choice. With regard to the prima facie evidence, what bearing do our senses play in discovering truth?

below. Witnesses said that the plane exploded into flames, instantly killing all aboard. Needless to say, my heart was ripped open and still bears a scar to this day.

I met Bob as a freshman in high school. We were classmates and became the best friends in the years that followed. I was originally drawn to Bob because he had something that was both puzzling to me and extremely different from what I had grown up with. He had a gentleness and a depth of understanding that was beyond me. Bob was light-years beyond his peers in maturity and insight into the human condition. He was a truly humble teenager who grew into a genuinely mature and authentic man. Bob saw people for who they were and looked deeper to who they could become. He made us all feel important and loved because he sincerely esteemed and respected everyone he met.

In a world that was anything but sincere or humble, Bob was a beacon of light in the darkness. As a high school boy, I was continually mystified by Bob's views of others and life in general. Didn't he know that some groups of people were to be avoided? Didn't he know that being cool meant doing certain things in certain ways? Yes, he knew, but somewhere along the way he had figured out what life was really about. He did not travel the path that most adolescent boys did, and yet his friends accepted him as a part of their group, no matter what that group was.

It was not until years later, after I had gone through many a fire and dark night of my own, that I finally began to understand the things Bob had said and done in high school. How was Bob the teenager able to comprehend so much of the human condition? How was he able to sincerely love and accept his friends when they were so far beneath him in intelligence and maturity? It often takes a retrospective look to understand things prospectively. There was never an agenda with Bob. He knew—I don't know how—that people are to be valued even when they're unattractive, insensitive, and foolish. His attitude towards life was to dignify everyone he met by being a servant when he was really a king. I don't think Bob ever viewed himself as a great person, much less a king. That, of course, is the nexus upon which humility turns, the king among paupers who is never fully aware of it. For me, only hindsight and a broader perspective on life have allowed me to fully recognize the sage who was once my friend Bob.

Simply put, the world needs more people like Bob. I often wonder, how much better life would be if Bob had lived his full life? How many more people would have been touched by his love and wisdom if he had lived to be thirty, forty, or even ninety years old? By living a life of profundity in a seemingly thoughtless world, he changed things for the better. Bob's attitude toward life is the topic of our next discussion.[122]

The first thing you must realize is that attitude is not an emotion, something that manifests itself when we are sad, happy, angry, and so on. Attitude is a chosen disposition of the mind (a mindset) that dictates how our emotions are internally viewed and acted upon. On a superficial level, attitude would be defined as intent. In the legal world, judges and attorneys are always looking for intent, or what the overall intent of a person's actions might have been. But on a deeper level, attitude is much more than intent. It's a mental state that a person assumes, incorporating elements of their personality, beliefs, and environment. In many circumstances, a person's attitude determines the mental structures they build for themselves, the choices they make, how they act, and how they view reality. In essence, a person will never be able to assess the truth of anything if their attitude about life is incorrect. And this is why some of the smartest and most affluent people on the planet are still the furthest from understanding any semblance of truth. By the same token, it's also why some of the poorest and most uneducated people living in the world can be so close to absolute reality and truth while the elites of the world are the furthest away.

For humans, attitude falls within two broad and general categories: humility and pride. A person is not usually defined by having one or the

---

122 In continuous memory of Bob Oakenell (March 17, 1962 – August 3, 1991). You walked in truth, dear friend. Thank you!

other. In other words, a person is not considered completely prideful or completely humble, but rather as a combination of both. If you recall the diagram from our discussion of free will and the set choice, there's a sliding scale between egocentric and altruistic choices. The more prideful a person becomes, the more egocentric their focus will be. Likewise, the more one moves towards an attitude of humility, the more altruistic one's choices become.

Believe it or not, there is a goal in life. That goal is to strive for balance. Not a state of being perfectly centered, as on the sliding scale just mentioned, but rather achieving a healthy balance in life. This means not only a healthy balance for one's self, but also for humanity as a whole and the earth as our home. The goal is a balanced Matrix, the basis and basics of life. A very real part of our heritage as human beings is to preserve the health (i.e., the balance) of life. It's a wonderful heritage to have. What higher goal is there? In absolute terms, humanity can claim no higher heritage than that of preserving life.[123] This is the basic stuff of life. Naturally, if we can't get the basics right, then we'll never achieve a healthy balance in anything.[124]

A person's attitude is pivotal to everything, from how we view ourselves, others, and the earth to how we systematize things through groups, organizations, governments, laws, information, and knowledge. Unfortunately, the behaviors that follow from egocentric attitudes cause dysfunction and unbalance in every aspect of life. It has been said and even taught in schools and colleges that things are the way they are because it's just human nature, i.e. it is what it is, and there's no use trying to change humanity. Let's assume for a moment that this last statement is

---

123 This is a simple universal truth that many past societies understood. It's truly ironic that today we would probably classify those societies as primitive.

124 We all share a moral responsibility to the planet we call home. The legitimacy of moral responsibility is an axiomatic truth to life. In modernity, we have this aversion to the whole concept of morals. Morals are viewed as manipulative tools that tell us what to do. Heaven forbid that we should bear any moral responsibility to each other and the planet we inhabit. Let's get past Enlightenment ways of thinking; morals are a truth and the way of life and without them we are lost.

true. If there's no higher good to strive for, if we cannot better ourselves and others, then where is life's meaning? Making a better video game, a smarter cell phone, or building a superior resume? Take a moment to think about this. Life's meaning is crushed in the quest to serve self (i.e., how wealthy can I be, how important can I become, how much popularity can I enjoy, how much love and respect can I claim?). The truth in this kind of meaning becomes nonmeaning, the very antithesis to everything good, admirable, and worthy in life.

Let's take a moment to define what we mean by attitudes of pride and humility, and then we'll discuss these things from a Socratic perspective.

Pride - An attitude of pride approaches information from the standpoint of superiority, or "I know more than you do." In a general way, this approach to information is what we've inherited from the Enlightenment (i.e., "Humans are the center and definer of everything.") We are the masters and stand above all things.

One can sense, almost immediately, a person who holds an attitude of pride. The prideful person will not listen to any view that conflicts with their own perceived objectives or reality. They are usually easily offended when their views are challenged in a meaningful and intelligent way. In order to protect their superior position, the prideful person will commonly go on the attack to discredit or slander the perceived offender. Any threat to a prideful person's information is an affront to who they perceive themselves to be; they have taken a personal position of pride rather than an objective and open position of humility. They have staked who they are as a person on how they have processed information in creating their own personal philosophy and worldview (i.e., their mental structure of reality is built upon the foundations of pride).

Humility - An attitude of humility approaches information from the standpoint of ignorance. A person with a humble attitude realizes their ignorance from the start. Contrast this approach to the person who is completely ignorant of their ignorance; the differences between the two attitudes are light-years apart. The humble person strives to understand, even if it means conflicts with currently held beliefs. In humility, understanding the information or the situation is far more important than satisfying one's own or another's self-interest.

With those definitions in mind, let's take a little Socratic test to explore these truths on a personal level.

How many abortions need to take place before we realize a human embryo is still a human regardless of its developmental stage? How many people today have any conviction about the millions who have died and continue to die in order for wealthy industrialized societies to live a lifestyle of excess? Does each person have a right to be sexually promiscuous while millions contract sexually transmitted diseases? Are broken families, a failing educational system, and systematic exposure to dysfunctional entertainment of any concern to myself or society?

Now, take a moment to assess your reaction to the questions just posed. Did any of the above questions offend you? What is your attitude toward those questions, and what can you actually learn from your reaction? The questions I asked are not unlike the questions Socrates asked of his countrymen. They are meant to test the individual on a number of different levels. To name just a few, the questions are meant to ascertain prejudice, emotional discipline, logical reasoning, and, of course, your overall attitude towards yourself, others, and life in general!

Our attitudes shape almost everything we experience in life. With the right attitude, we can change our thinking styles, mature in wisdom, and experience a life we never knew existed. Reflect on this for a moment. Absolutely nothing happens in a vacuum. Our actions or inactions always affect other people, either now or in the future. Whether we process information as a Style II or III thinkers, whether our perceptions of reality are from Enlightenment or Rationalist perspectives, or whether our attitudes are based primarily on humility or pride, we always have an impact on the world, for good or bad, that ultimately confirms or denies the ultimate truths and absolute realities of life.

So, what conclusions can we draw? Here's one among legions: free will comes at a price, and it would appear that price is pinned exclusively to our moral obligations to each other.[125]

---

125 Poverty, war, disease, envy, hatred, corruption, murder, pollution, climate change, environmental destruction, and suffering can all be traced back to the set choices of people. To be sure, there are viruses and bacteria, earthquakes, hurricanes, and diseases that do not have human origins and still cause human suffering, but at least 80 percent of the world's pain, suffering, and hardship is directly associated with the self-serving choices humanity makes, choices made almost exclusively with egocentric intentions and motives in mind. What does that say about who we are as individuals and a society?

– CONCLUDING REMARKS –

"If this be hard to understand, it is as the simple, absolute truth is hard to understand. It may be centuries of ages before a man comes to see a truth; ages of strife, of effort, of aspiration. To see a truth, to know what it is, to understand it, and to love it, are all one. There is many a motion towards it, many a dim longing for it as an unknown need before at length the eyes come awake, and the darkness of the dreamful night yields to the light of the sun of truth."

—George MacDonald

The absolute truth of anything will remain hidden until we can truly see who we are, fully exposed, fully measured, in the consuming light of the truth. Until we see not just the things we have intentionally hidden and kept secret, but also those things that remain hidden from our conscious selves, understanding universal truth and reality will always be beyond our grasp. Nothing less than an epiphany must occur, a total acceptance of who we really are and are not, and of who we are meant to be in light of that truth. It is unlikely that you can reach such a place without feeling a great loss within yourself. For most of us, it will be the realization that we are not the people we believe ourselves to be. When the illusions come crashing down and we are left naked and humbled before the absolute truth of ourselves, it will be then, and only then, that we can comprehend where we are in relationship to absolute truth and reality. This is the starting point, the beginning of a new life that puts away the old self and begins anew. Authenticity, humility, and wisdom go hand in hand. Where there's no authenticity and no humility, there's no wisdom or truth to be gained. But, where authenticity and true humility are present, wisdom is born and absolute truth is revealed to the individual.

It is truly ironic that the smartest species on the planet are also the ones destroying it. Why is that? It's because we are inauthentic with ourselves, with others, and the reality that gives us life. We have become disconnected from all that sustains us. While it is exciting to view all the technological and scientific changes that are taking place daily, we still need to strive for those moral changes, i.e. those set choices, that will allow us to continue to live good and productive lives for millennia to come. Unfortunately (in modernity), we are not living for the long-run. If we were, things would be vastly different on our planet. This is an incredibly poignant fact facing us today. We are the only animal on the planet that progressively works in a regressive fashion, and that must stop if we are to save ourselves.

All of our conversations so far have been to prepare us to dive deeper still. It will be at this place, the great abyss, the divide between categorical and quantum thinking, that most people will surface for air and refuse to dive any deeper—if it hasn't happened already. This is tough stuff to handle, especially when it deals with you and me and not somebody else. If I were to encourage you in any way, it would be to withhold any final judgments based solely on the conversations we've had so far. The differences between linear and quantum realities are vast, and it takes time to understand them. Hang in there! You now have a unique opportunity to learn something that comes from outside your knowledge base. The key question is, what will you do with it?

A fool's response is to always jettison any information that conflicts with his or her worldview. Wisdom dictates that judgment should always be slow in coming. Even those things we deem inconsequential or ridiculous are opportunities to learn. The bright side is that now we *know*. We always desire to know, and now we know many things that were not self-evident before we started our discussions.

Each of us has the ability to go beyond our present position. I have witnessed over and over again as people (myself included) made great strides in overcoming the past and using the future as a springboard to new beginnings. Most people want to do the right thing; most want to be fair to their neighbors, whether they're next door or in another country. And most people do have a deep seated need to be loved and to love others. These desires are not only good things to strive for, they just happen to be the best things life has to offer. Each of us has an indomitable spirit that strives to be let loose, a spirit that wants so desperately to have a noble purpose. The good news is we can have meaning in life! We do not have to take what the media and popular culture have to offer (self-induldgence). We can change it all for the better if each of us individually wants to better ourselves from this point forward. Certainly the road will be rough, but not impossible. It's hard times and reaching for seemingly unattainable goals that make us stronger and better people.

There's an old African saying I heard many years ago that goes something like this: "When is the best time to plant a tree? Yesterday. When is the next best time to plant a tree? Today!" You can plant that tree today, so that at the end of your life you can look back and say, that's when I began anew, that's when my life started growing beyond mediocrity, that's when my life began to shelter those in need, and that's when the truth began to fully live within me. What a legacy to leave, and what a gift to offer up to humanity! It's here, in taking action, that the ultimate truth comes alive. Don't miss this opportunity to live for a greater good and a greater cause. I know you can do it!

– EPILOGUE –

*"If by supporting the rights of mankind, and of
invincible truth, I shall contribute to save from the agonies of death one
unfortunate victim of tyranny, or ignorance, equally fatal, his blessings
and years of transport shall be sufficient consolation to me for the
contempt of all mankind."*

—Marquis Beccaria

A big storm was blowing in from the east. The winds were strong, leaves were blowing all around, and stammering bits of rain flew sideways across the muddy path. At times the wind was a deafening roar, a force to be reckoned with, and then it would subside, only to rebuild its strength for another thrust. You could hear it coming through the forest, a tsunami borne upon the treetops, building strength and energy as it approached my spot. Then, like a huge breaker cast upon rocky cliffs, it would hammer me as I struggled on.

As I neared the village, I saw a dark shape beneath a tree. At first I thought it was an animal of some sort, but as I drew near it became obvious that it was a person. With head bowed and knees drawn up, the person appeared to be meditating or sleeping. The contrast between the person and the tree was striking. The tree was dead, its trunk and branches stripped and white as bones. The person was dark. Dark trousers and a dark hooded sweatshirt pulled over the head. In many ways, the person seemed to be a part of the tree, and yet one was as distinct from the other as black on white.

As I walked by, I noticed how small and insignificant this person seemed. The wind and rain flew around person and tree, but neither seemed to be affected. In that moment, something within me was speaking, something I could not fully hear or understand, so I made a mental note to return to the scene after I made my visit to the village.

When I entered the village proper, I headed straight for the elder's hut. Upon entering I noticed someone lying on the earthen floor, completely covered from head to foot. When the elder heard me enter his home, he greeted me as an honored guest and welcomed me back from a long absence. When I gestured to the person on the floor, he told me that it was Merri's grandmother, who had just died of malaria.

When I first met Merri, he was five years old. He was all smiles, a nuclear power pack of energy. Merri had raven-black hair that was soft and light. His eyes were dark, but with the luster of polished obsidian. He was slender as a reed, but strong as iron. I'd never seen a five-year-old boy who could climb trees as fast or wear his parents down as quickly as Merri could. When Merri saw me, he never failed to give me a hug. Always, it seemed, he was grabbing my hand and dragging me off to his next adventure. Merri's faith in the future was contagious; he was quite the special little man.

When you looked into Merri's eyes there were many things to be seen. They sparkled in a happy, playful sort of way, and by this special glint you could tell that Merri loved life and wanted to experience it all. The second thing I noticed about Merri's eyes was the way they looked at you. He would study you as a jeweler studies a diamond. When he caught you looking back at him, his eyes would lock onto yours, penetrating deep, seemingly looking for something beyond the surface. With me, Merri found that something; I loved Merri with all my heart. Somehow I knew we had a bond meant to last a lifetime.

The elder touched my arm and brought my memories back to the present. He related a story to me, one that had transpired in the two years I was away. Shortly after I left the village, Merri's mother died of a disease that is common in that region of the world. Before long, Merri's father died of the same disease. With no parents, Merri and his brother went to live with their grandmother. Because of the extreme poverty in the area, food was scarce, even more so since Merri's father was no longer around to work for the family. So it was left to Merri and his brother to work to feed their grandmother and themselves. All three of them were reduced to living on one meal a day and sometimes had nothing to eat at all. A year after the boys moved in with their grandmother, Merri's brother died of malnutrition. Not too many months thereafter, Merri's grandmother died of malaria. Merri was now alone in the world.

When I asked the elder what had become of Merri, he walked me outside his hut and pointed to the dead tree, the one I had passed before entering the village. The elder said the person sitting at the base of the tree was Merri. I thanked him for telling me about the circumstances, and then I excused myself and walked to where Merri and the tree sat unmoving while the storm raged about them.

The rain was coming down in sheets now. The path I had traveled was now a strong-flowing brown river. When I reached the tree, I knelt down and touched Merri on the shoulder. Merri turned his head towards me, and when recognition dawned, he leaped into my arms. The embrace lasted forever. Tears of happiness turned to tears of intense grief. I eventually pulled Merri away and held him at arm's length. As I looked at him, Merri's eyes were the same as I remembered them, reflective, searching, and still smiling. The glint in his eyes and the smile on his face said it all: he loved me still, his indomitable spirit was amazing.

Merri died two weeks later. Merri had many cards stacked against him, so it was hard to tell what exactly killed him, but malnutrition and a weakened immune system were the primary culprits. When I returned to the United States, I mentioned Merri's death to a friend—I'll call him Sam. Sam holds an MBA from a prestigious university and has always worked for large corporations. Not surprisingly, he is extremely smart, follows all the latest developments around the world, and constantly tracks the stock market. After I told Sam about Merri, he said, "I'm sorry to hear about that, but those kinds of things will change as soon as free markets start expanding worldwide." I just stared at Sam, unable to speak. Then he added, "Oh, by the way, almost forgot, I have two tickets to the football game tonight. You interested in going?"

A Forecast

Everyone seems to love forecasts. Usually trying to forecast anything, especially the further into the future the forecast goes, is pure folly. Because of the shear magnitude of the variables involved, predicting the future becomes an exercise in grand speculation. But even so, we all like predictions that have the possibility of becoming true even if we can not know with certainty their ultimate fulfillment. With that said, let's talk about possible predictions associated with the discussions we have had.

Since this book was written and based upon universal truths and the underlying reality that supports those truths, we can have certain expectations for the future, those expectations being two fold:

1) First, that the future is found in present and past circumstances, e.g. that cause and effect relationships have always been found in past, present, and future events. Likewise, historical evidence can and often does produce facts. Facts that can be used to determine cause and effect associations for the future.

2) Second, there are two fundamental laws (universal truths) governing free will that will remain unchanged in the future, those being altruistic and egocentric choices (refer back to the 'Set Choice' discussion).

With those expectations in mind, the following predictions should hold true:

- The self-centered "me" mentality (Style II and III thinking) will grow exponentially as the population of the earth grows, hence, this will manifest itself in the explosive growth of the entertainment industry, it all its forms.

- When it is finally realized (i.e. fully conceptualized) that science and technology creates as many problems as it solves, there will be a great moral upheaval across the planet as the earth nears its irrevocable collapse.

- Due to global bifurcation, natural disasters will increase, e.g. earthquakes, hurricanes, tornados, etc.

- And, due to monetary security concerns, a one world system will be established to control for potential monetary risk. This world system will be an economic system based upon capitalistic principles.

The really great unknown to predictions is the human equation. Human beings have an amazing ability to change behaviors when faced with death. The question is never what will be, but what will we become? Will humanity, us, you and I, embrace our short comings and ignorance as fundamental parts of existence? Can each of us change enough individually to make a difference globally? Can we make room for the new by getting rid of the old? Can we replace pride with humility, selfishness with servant-hood, and freedoms with sacrifice?

That is the quintessential question for modernity. Ironically, that question may even now be too late to ask. Already, we could be past the point of no return. Is the die already cast? No one really knows for sure. Should we try to change our thinking and ways of behaving to align with reality? Of course we should. And by doing so now, we may just be able to dodge a really black prediction and an even darker future.

Here ends our discussions on the Matrix of Truth and the Quantum of Reality. Book 2 will dive even deeper into those quintessential questions of life. It's the questions that always compel us, it's our need to know that drives us, and it's the answers that can inspire and change us if we let them. Until then, continue to learn, take actions small and large, and grow in wisdom and humility, my friend.

– GLOSSARY –

– A –

acquired immunodeficiency syndrome (AIDS): an immune-system disease caused by the human immunodeficiency virus (HIV) and transmitted primarily through unprotected sexual intercourse and use of contaminated needles. In the United States, AIDS was first identified as an epidemic in 1981 after a sudden increase in cases directly linked to promiscuous behavior in homosexual male communities.

anthropic principle: the cosmological consideration of why and how the universe is capable of supporting life, particularly human life. Is the universe the way it is because that is the way that humans perceive it, or do humans perceive it that way because they are a product of the universe? In other words, is our universe unique and single in the fact that everything is fine tuned to the point of allowing intelligent life? Do the dimensionless physical constants explain the existence of life (carbon based life), or do they necessitate or even require carbon based life to possess a consciousness to perceive a reality beyond itself?

anthropomorphic: giving human qualities to nonhuman things. For example, alien beings are depicted as creatures that walk upright and have heads and appendages similar to those of humans. We ascribe human features to unknown beings because we don't know what an alien would look like. This applies to Greek and Roman gods as well.

antimatter: Matter composed of antiparticles, e.g., particles that differ in their electrical properties/charges but are equal to each other by mass.

a posteriori analytics: knowledge that is learned after the fact (i.e., from past events or history) expressed as an empirical fact unknowable by reason alone.

a priori analytics: knowledge based entirely in study or analysis of a subject instead of the use of prior knowledge or experience.

arrow of time: a timeline that moves in only one direction, from past to present and into the future. The arrow of time has direct links to the Second Law of Thermodynamics (irreversibility) and the measure of disorder in a system, entropy.

attribution asymmetry: a person's ability to ascribe their achievements to their own abilities, genius, or physical talents and their weaknesses or failures to external circumstances. A classic example is the person who blames others for their failings while at the same time extolling their own intellect, drive, and self-importance for their achievements.

– B –

Berkeley, George (1685–1753): George Berkeley was a philosopher who established (in his own mind) that objective reality and universal truths are subjective in nature, based on each individual's perceptions and/or experience. In Berkeley's view, material objects cannot exist in and of themselves without a mind to perceive them.

bifurcation (Global): A global system breakdown which is predicated by a measurable increases in atypical variations to archetypal situations, i.e. increase in global warming, earth quakes, droughts, tsunamis, hurricanes, disease, etc. One could also extrapolate bifurcated anomalies on a social scale, i.e. wars, mass shootings, suicides, child abuse, etc.

blank slate (tabula rasa): the theory that infants are born with a blank mind, that is, with no preexisting, a priori knowledge on which to draw in order to make decisions or understand their surroundings.

- C -

Cantor, Georg (1845–1918): the German mathematician who invented set theory and transfinite numbers, Cantor is primarily known for his mathematical proof of infinity, the infinite set.

cartesian plane: a four-coordinate mathematical system that uses the center point of the plane (the origin) as a foundational reference point in order to approximate measurements in an infinite space. The Cartesian plane was invented by René Descartes.

casimir effect: a quantum term for something being where nothing is supposed to be. In other words, what we once thought was a vacuum or empty space is now known to contain something, in this case vacuum fluctuations or quantum jitters. This fact was proved by the Dutch physicist Hendrik Casimir in 1948.

closed system: a system that is seemingly self-contained and operates independently. For example, the earth could be considered a closed system in that it creates and operates its own environment. From a biological perspective, the human body could be considered a closed system in that the body functions and labors to maintain itself. On a larger scale, the universe could be considered a closed system in that it apparently operates within certain constraints. Closed systems seemingly operate as finite realities. *See also* open system.

cognitive analysis: a way to measure thinking abilities and styles.

cognitive dissonance: the discomfort experienced when one internally held belief comes into conflict with an opposing belief, where both beliefs are internalized within the same individual. One must decide which belief is correct in order to avoid the discomfort (the dissonance) of holding both positions/beliefs at the same time.

coherence: in physics - two waves having a fixed phase/frequency relationship where phase differences between waves is constant, holding a quality of systematic consistency. In reality - the truth of any proposition must cohere (fit) within the parameters of any universal truth.

comparative methods: a research technique used to compare differences and similarities within systems, principally cause-and-effect relationships and how those relationships build upon the goals of determining reality and universal truth.

compressed energy: energy stored in a compressed state, as in a spring or air cylinder.

confirmation bias: the tendency to favor information that confirms one's point of view or belief system.

connected coherence: a fixed and connected relationship between entities, substances, and/or deductive systems.

content analysis: the systematic content/element analysis of communications to determine the objective reality and universal truths associated with such.

convergent systemization (correlational triangulation): the confirmation of information by independent, dissimilar, and diverse sources of information.

Copernicus, Nicolaus (1473–1543): the mathematician and astronomer who proposed that the sun, not the earth, was in the center of the universe. Until that time, humanity believed it occupied a special place in the universe, and thus, everything must revolve around the earth and man.

cosmological constant (Einstein's version): Einstein proposed the cosmological constant to make his equations for general relativity reflect a static universe instead of the expanding universe that his equations ultimately predicted. The adjustment was a reaction to a bias in the physics community at the time, the belief that the universe was fixed and unchanging.

cosmological constant (current version): Today the cosmological constant is believed to be the dark energy that comprises 70 percent of the whole of the mass/energy equivalent.

– D –

dark energy: a form of energy that infuses 70 percent of space.

decoherence: the process by which a system's behavior changes from that which can be explained by quantum mechanics to that which can be explained by classical physics, i.e. the closed system.

deductive logic: the process of reasoning from a premise to reach a logical conclusion (e.g., if the premises are true, the conclusion is, out of necessity, true).

Democritus (460–370 BC): the philosopher who believed in an eternal deterministic chain of events coalescing in one final outcome for the future.

Descartes, René (1596–1650): the rationalist philosopher and mathematician whose deduction "I think, therefore I am" and the Cartesian coordinate system made him a major contributor to rationalistic methods of reasoning.

dialectics: the Socratic method of reasoning through dialogue using deductive methods to arrive at a reasonable conclusion.

Diogenes (412–323 BC): the Greek philosopher who maintained that modern society was a falsity in its complexity and that morality meant a return to simplicity. Diogenes famously wrote, "I am a citizen of the world."

distal perspective: the pure cause of something yielding a universal truth. Distal perspective is a rationalist approach to determining reality.

domino effect: the cumulative effect produced when one event sets off a chain reaction.

doppler effect: A change in the frequency of a wave moving relative to its source resulting from the motion of the source and the medium in which the wave travels.

dualism: the quality of having a twofold nature. True dualities are pure truths in that they establish the baseline for universal truths.

-E-

empiricism: a theory of knowledge in which direct sensory experience is the only avenue for the discovery of new knowledge and evidence. *See also* positivism.

Enlightenment (the Age of Reason): the eighteenth-century philosophical movement that gave life to modern thinking, rejecting rationalism in favor of empiricism.

entanglement: the nonlocalized connection of shared states of indefinite being between particles. This is that part of the Matrix within the open system that spills over into and becomes entangled with our closed system. It is the infinite/open system which is tied to the finite/closed system.

entropy: the trend of the universe to move from an ordered state of being (low entropy) to a disordered state of being (high entropy). Entropy is also a function of the arrow of time and is considered with the thermodynamic variables, as temperature, pressure, or composition that is a measure of the energy that is not available for work during a thermodynamic process. In a closed system, organization eventually evolves toward a state of high entropy.

Epicureanism: a philosophy that believes hedonistic pleasure is the greatest good. This philosophy was adopted and generally espoused during the Enlightenment period and has direct ties to how we think and act today.

epistemology: the study of the origins, limits, and validity of knowledge.

ethical examination: a study or examination that looks at the consequences of what is considered good or bad for society.

ethnography: qualitative research designed to examine cultural norms and mores.

Euclidean geometry: the mathematical study of flat space using a finite system of postulates and axioms. Euclidean geometry remained unchallenged for more than two thousand years and was deemed irrefutable as a universal truth. It was not until the nineteenth century that questions arose concerning Euclid's simplistic approach to spaces. *See also* non-Euclidean geometry.

– F –

falsifiability examination: Something that is capable of being tested as either falsifiable or verifiable.

Feynman, Richard (1918–1988): the physicist who was instrumental in completing the concept and model for the path integral formulation of quantum mechanics.

fine-structure constant: the dimensionless electromagnetic interaction between particles. Physicists know this dimensionless quantity to be a fundamental physical constant that is universal in both nature and time.

fluctuating energy fields: rapid energy fluctuations in particles or fields that generate the vacuum or background energy within the universe known as dark energy.

Friedman, Jerome (1930-): the American physicist who first discovered evidence that protons had an internal structure, later to be known as quarks.

– G –

genovese syndrome: a social psychological phenomenon in which individuals do not offer help when other bystanders are available to help.

global bifurcation: a global system breakdown that is predicated by measurable increases in atypical variations to archetypal situations, such as an intensification of global warming, earthquakes, droughts, tsunamis, hurricanes, or disease. Bifurcated anomalies might also be extrapolated on a social scale, in forms such as wars, mass shootings, and increased rates of suicide or child abuse.

Gödel's incompleteness theorems: two mathematical logic theorems showing that consistent and complete mathematical axioms are impossible to have.

- H -

heuristics: the mental shortcut used when faced with complexity. Reductionism would be an example of a heuristic way of looking at things.

historiography: the examination of how history is interpreted and recorded.

hermeneutics: the study of the methodological principles of interpretation.

Hubble, Edwin Powell (1889–1953): the American astronomer and cosmologist known for his discovery of the recessional velocity of our galaxy and the subsequent implication that our galaxy, and possibly our universe, is expanding from a low state of entropy to a high state of entropy.

human papillomavirus (HPV): a sexually transmitted virus that causes lesions, warts, and various precancers and cancers of the skin lining the lower genital tract, anus and mouth. It is the most common sexually transmitted virus spread almost exclusively by promiscuous sexual practices.

humanistic constant: the human drive for self-actualization and self-realization. Its focus is always self-discovery.

– I –

idealism: a theory that only the perceptible is real. Therefore, reality is entirely a construct of the mind and nothing more. Knowing anything outside of one's mental abilities is impossible.

inductive logic: a system of evaluative support that evaluates general propositions and reduces them to specific examples or findings.

inductive coevolution: a system of analysis that tracks the evolution of change by way of inductive reasoning.

informational influence: seeking informational guidance from others as a result of the desire to conform to the reality of others, instead of conforming to the realities of universal truths. Also called social proof.

informational wave (or stream) energy: a particle wave or stream that carries with it information in the form of quantum instructions having to do with energy, mass, particle, and so on.

interpretive deconstruction: deductive analytics used to remove interpretive or subjective truths from qualitative studies or conclusions.

Irenaeus (142–202 BC): the bishop of Lyons who wrote prolifically and philosophically about epistemological concerns related to Gnosticism.

– J –

Jamison, Kay Redfield (1946–): an American clinical psychologist and professor of psychiatry at the Johns Hopkins University School of Medicine, specializing in bipolar disorders.

- L -

Leibniz, Gottfried Wilhelm (1646–1716): a German mathematician and philosopher who used rationalism in the discovery of truth. Leibniz also developed infinitesimal calculus and the Leibniz notational system for calculus.

linearity (linear): consisting of a straight line, as in a straight line of progressive thoughts, a straight line of being (the arrow of time within a closed system), a straight line of reasoning (e.g., classical mathematics and science, regressive forms of analysis).

Lucretius (ca. 94–51BC): a Roman philosopher who espoused Epicureanism.

- M -

macrosystem: a larger system in which smaller microsystems exist. For example, the ocean would be considered a macrosystem, while a coral reef would be considered a microsystem. Likewise, our galaxy is a macrosystem containing the microsystem of earth.

mass / energy equivalence $(E = mc^2)$: the equation derived from Einstein's theory of special relativity, expressing the relationship between the mass and energy of an object(s) where E is the energy of the object in joules, m is its relativistic mass in kilograms, and c is the speed of light in meters per second.

Matrix of Life: The macro- and microsystems that connect all being (i.e., ultimate truth with essential reality, the infinite with the finite, the physical with the mental, energy with information, etc.). The Matrix of Life is the one all-encompassing system that unifies all systems.

matrix analysis (connected coherence): the study of the characteristics of objects as they relate to macro- and micro-level systems within the Matrix of Life. *See also* connected coherence.

microsystems: a system that exists inside another system (i.e. the macrosystem).

moral disengagement: a term from social psychology describing the exemption of one's self from compliance with moral or ethical standards in a particular situation or context.

moral examination: the quantitative and qualitative analysis of moral and ethical principles as related to the health and welfare of the whole (i.e., the Matrix of Life), including environmental and societal concerns.

- N -

nanoparticle: a microscopic particle with dimensions less than 100 nanometers.

Nash, John Jr. (1928–): the American mathematician who suffered from paranoid schizophrenia while pursuing mathematical research into differential equations, game theory, and differential geometry. His work has provided insights into the complex systems that govern the dynamics of everyday life.

Newton's constant: the universal closed system constant that relates force to mass to distance in Newton's law of gravitation. It is the force of the gravitational attraction between two bodies to their masses and their distance from each other.

Newtonian perspective/view : this view of life is firmly grounded in the obvious. In other words, if something appears to conform to certain physical laws or even to social customs and standards (i.e. a majority opinion, custom, or standard), then the law or standard is said to fit a classical role or a Newtonian perspective of life.

network analysis (stratification): the study and classification of symmetric and asymmetric relationships as viewed from a holistic network. The Internet is one example of a holistic network, and groups and organizations are examples of social networks/stratification. *See also* stratification.

neutrino: a neutral subatomic particle with a mass close to zero and half-integral spin, rarely reacting with normal matter and thus having the ability to pass through matter.

non-Euclidean geometry: the mathematical study of space with more than three dimensions.

nonlocality: the theory, closely tied to quantum entanglement, that nonlocal objects can interact (communicate) with each other even when separated by space and time

- O -

obedience, unconditional: When one chooses to obey another's demands without placing any stipulations on the demands.

Ockham's razor: the concept that the simplest explanation is probably the best explanation, attributed to William of Ockham. *See also* parsimony, law of.

open system: a system that surrounds and infuses closed systems with energy (sustaining life in all its forms) and information (knowledge in all its forms), both of which are necessary for closed system dynamics. Open systems are logically deduced by way of scientific discovery (e.g., as consequences of general relativity and quantum mechanics) or by way of mind experiments (e.g., theoretical physics, mathematical relationships, cosmological induction, and philosophical reasoning). As such, open systems are believed to be infinite and eternal in nature and form; hence, the origins of transcendent truth and objective reality have their essences firmly established within the matrices of an open system.

- P -

Paine, Thomas (1737–1809): the English-born political activist who promulgated the ideals of the Enlightenment to the American colonies in the eighteenth century.

Pandora's box: a metaphorical box that releases havoc into the world as a result of unwise decisions.

Pareto efficiency: a state at which no further change can make one person better off without making another person worse off.

parity inversion: the mirror symmetry of a constant (e.g., time reversal, conjugation, and space inversion). Symmetries are tied to conservation laws due to their reciprocal properties.

parsimony, law of: the concept that the simpler an explanation is, the better it is. Its prime objective is to remove as many assumptions as possible before coming to any conclusions. Parsimony is used as a general rule in science when developing models and theories. See also Ockham's razor.

perceptional wisdom: the ability to understand the complex, to perceive the unknown, and to act accordingly.

phenomenology: the study of human self-awareness as it relates to experiences and/or objects.

phase angle: a cycle expressed in time. A shift in the phase angle also means a shift in time.

Piaget, Jean (1896–1980): the Swiss developmental psychologist who theorized that knowledge is constructed and built upon only after we are born, not before.

Planck length: approximately 1.616×10^{-33} centimeters.

postdiction inquiry: the hypothesis that some unknown event happened in the past; the act of making a prediction about the past in order to understand present or future conditions.

positivism: the theory that reality is defined by what humans can directly experience through empirical data by way of direct experience, observation, and direct measurements. Contrast with realism.

Post, Emil (1897-1954): the American mathematician who suffered from clinical depression while contributing valuable mathematical methodologies for computing logic, recursion and polyadic theory.

pragmatism: a compromise between universal truth and the expediency of the situation. Pragmatism is also known as situational ethics due to its focus on circumstances and self-interest versus universal truths.

THE MATRIX OF TRUTH

proximal perspective: the prospect of gaining knowledge by being situated next to or near the point of universal truth. Rational inquiry from a proximal perspective can and often does yield universal truths.

Ptolemy, Claudius (ca. 90–ca. 170): the Roman geographer and mathematician who believed the earth was the center of the universe.

pure number: a number quantity that does not have physicality.

– Q –

quantum entanglement: that part of an open system that is in a collective indefinite state of shared commonality (e.g., electrons, photons, etc.) until it is subjected to (measured from) a closed system state. *See* closed system and open system.

quantum geometry: mathematical constructs and/or algebraic systems used in quantum mechanics to describe the physical phenomena happening at short distances that are comparable to the Planck Length.

quantum mechanics: the branch of science that deals with the structures and interactions of subatomic particles as they relate to matter.

quantum physics: the branch of science that deals with discrete energy units called quanta on a subatomic level. Quantum physics addresses properties of atoms, nuclei, subatomic particles and light.

– R –

rarefaction (energy): Usually the reduction of an item's energy or density over space at a single point in time, or a reduction of density through time.

rationalism: the methodology in which truth is derived by use of intellect alone, as opposed to empirical analysis. *See also* <u>realism</u>.

realism (e.g. Scientific Realism / Philosophical Realism): a methodology that states reality exists beyond the perceptions and constructs of man. Leibniz's concepts of existence, Einstein's theories of relativity (both general and special), and Heisenberg's quantum mechanics are all examples of a reality that stands beyond man's experience or observation (i.e., a deduced reality). Contrast with <u>positivism</u>.

recursive analysis: a system of assigning values to an infinite number of variables, which in turn can then be used in a finite set of rules. *See* also <u>parsimony, law of</u>.

redshift: the production of longer wavelengths as an object moves away from the earth. Redshift was used to determine the expansion of the universe.

reductionism: a procedure or process that reduces complex data to simplified terms. This mental construct or tool is used when complexity is overwhelming. *See also* <u>heuristics</u>.

regressive analysis: a method for approximating complex relationships and variables in order to arrive at approximated answers.

Renaissance: the rebirth of art, science, literature, and culture in Europe, beginning in fourteenth-century Italy and lasting until the seventeenth century.

repetition bias: the effect in which people tend to favor or believe information that has been repeated over and over again.

- S -

Schrödinger, Erwin (1887–1961): the Austrian physicist who created the mind experiment known as Schrödinger's cat. Schrödinger also developed matrix mechanics and wave functions.

simultaneity: simultaneity and superposition operate hand in hand. One cannot be defined by itself as they both define each other. The simplest definition of simultaneity states that an object (in quantum mechanics, a particle) occupies more than one position at the same time, e.g. simultaneously. In quantum mechanics, superposition is simultaneity. This phenomenon is directly associated with the open system that surrounds and is outside of the closed system in which we live. *See* also <u>wave function</u> and <u>entanglement</u>.

social conformity: thinking and acting in such a way as to be accepted by a person, group, corporate entity, a government, or other cultural entity. Thinking and/or action taken in accordance with some specified standard or authority in an effort to be socially acceptable to others.

space–time dimension: a four-dimensional continuum that exists within a closed system made up of time and space.

stratification: the study and classification of symmetric and asymmetric relationships as viewed from a holistic network. The Internet is one example of a holistic network, and groups and organizations would be examples of stratification. *See also* <u>network analysis</u>.

superposition: the property of a particle occupying all quantum states simultaneously. Closely associated with simultaneity, in which two or more events can happen at the same time using the same point of reference. *See also* <u>simultaneity</u>.

- T -

thermodynamics, first law of: the energy contained within a closed system remains constant, closely associated with the law of conservation of energy.

thought experiment: a demonstration of pure rationalism based solely on what the mind can comprehend that either clarifies or demonstrates the consequences of objective reality and absolute truth.

time dilation: the slowing down of time by motion.

- U -

uncertainty principle: the impossibility of measuring the velocity and the position of a subatomic particle simultaneously.

- V -

vacuum fluctuations: a spontaneous fluctuation of energy within a vacuum that defies the axiom of the conservation of energy. Used to explain virtual particles that are associated with the uncertainty principle. In quantum mechanics there exists (within a vacuum), a spontaneous source of energy fluctuation. This spontaneous source is in direct conflict to the axiom of the conservation of energy, but is mathematically used in order to explain virtual particles as associated with the uncertainty principal.

- W -

wave function: the probability of a particle's amplitude (i.e., how it behaves in an open system state). *See also* open system.

westernized methodologies: thinking and teaching styles that have been passed down from the Greco-Roman, Judeo-Christian, and the Enlightenment periods. For the most part, the Western world consists of Europe, North America, and the Latinized countries. Today, most of the world is educated under westernized doctrines of thought and action.

– ABOUT THE AUTHOR –

"I refuse to join any club that would have me as a member."
Groucho Marx

On many different levels, the above quote by Groucho Marx pretty much sums up who I am, who I'm not, and where I am in this glorious adventure called life. There are legions of things I could say about myself (the usual successes that most authors list; education, awards, etc.), but in the eclectic spirit of this book, I'll provide the reader with a very personal vignette of something that has literally defined my entire life.

Life is just full of nifty surprises, some of them quite pleasant to experience and others quite debilitating. For most of my life, the latter has held true. You see, I deal with a mental illness and for a multitude of reasons, I simply call this disease the beast.

For quite a long time I kept my mental condition hidden, even from myself (subconscious mind to the rescue). Such are the denials that most people with mental disorders have. In hindsight, it's actually quite comical thinking back to all the ways I tried to control and hide the beast that was dragging me through life kicking and screaming. It was not until my forties that I came out of the closet, so to speak, and fessed up to having a mental disability. Funny thing, I found out that everyone already knew I was mentally deranged before I did. I wonder how I missed that one?

The reality of the disease is that it's a shadowy monster that takes over and controls much of my life. Life for me has come in small spurts of light (happiness), while death has been lived in colossal swaths of darkness (intense depressions). In other words, I've spent most of my adult years living in a mental hell, medications notwithstanding. So, you may be wondering, how has this mental condition affected myself and those around me?

My cognitive abilities tend to be chaotic. At times I process information as a person with dementia, struggling just to think clearly enough to remember tasks I've done hundreds of times before. Sometimes I have profoundly lucid moments of insight, clarity, and deduction, but I'm finding even those times have grown less and less with time. Thus, during those times when I am thinking clearly, my research work proceeds at breakneck speeds as I know it will not be long before the beast returns and my depression takes me again.

On the social side of things, I tend to keep my personal interactions short-lived. Since my moods and depressions are unpredictable, people don't' tend to tolerate me for any length of time (fully understandable). Because of this, books tend to be surrogate friends, journaling ends up being my emotional release valve, and research work becomes the passion for understanding not only my disease but so too the world around me. For all intents and purposes, I've lived as a quasi recluse. Many would constitute this as a strange life, but it's one that has worked for me. It helps me to cope with the world, but equally true, it also helps the world to cope with me. In many regards, I consider it a win/win situation.

And on a final note of interest, there are two main subject areas that have always fascinated me: the first is science with all its attendant physical, social, and theoretical attachments, and history with its records, cultures, and applicable lessons to be learned. All of it: science, math, philosophy, religion, law, government, psychology, etc. have fascinated me since I could read, and it is to these many subjects that I have dedicated the greater part of my life.

www.ingramcontent.com/pod-product-compliance
Lightning Source LLC
Chambersburg PA
CBHW070805050426
42452CB00011B/1899